The Science of Cooking

Springer

*Berlin
Heidelberg
New York
Barcelona
Hong Kong
London
Milan
Paris
Singapore
Tokyo*

Peter Barham

The Science of Cooking

 Springer

Dr. Peter Barham
2 Cotham Place, Trelawney Road
BS6 6QS Bristol
United Kingdom

ISBN 3-540-67466-7 Springer-Verlag Berlin Heidelberg New York

Library of Congress Cataloging-in-Publication Data

Barham, Peter, 1950 –
   The science of cooking / Peter Barham.
      p. cm.
   Includes bibliographical references and index.
   ISBN 3540674667 (alk. paper)
      1. Cookery. I. Title.
   TX651 .B147 2000
   641.5--dc21

                               00-059559

Springer-Verlag is a company in the BertelsmannSpringer publishing group.
© Springer-Verlag Berlin Heidelberg 2001
Printed in Germany

Production editor:
Cover design:
Typesetting: Fotosatz-Service Köhler GmbH, Würzburg

SPIN: 10679403   24/3134 ra – 5 4 3 2 1 0   –   Printed on acid-free paper

Contents

Acknowledgements

There are many colleagues and friends who have helped and encouraged me in the writing of this book. First, I must express my gratitude to Sue Pringle, who is largely responsible for persuading me to start giving public lecture demonstrations on the science of cooking. Secondly, I must thank all the delegates at the International Workshops on Molecular and Physical Gastronomy in Erice over the last few years. I was particularly inspired by Nicholas Kurti, who was the driving force behind the workshops. The many discussions at Erice (and elsewhere) with Leslie Forbes, Tony Blake, Herve This and Len Fisher have wasted many pleasant hours and often clarified ideas in my mind and provided ample motivation to get on with the writing. I am also indebted to Ramon Farthing for explaining his cooking techniques patiently to me.

However, the person who has most influenced me and has given the greatest assistance, by suffering the testing of all the recipes, reading and re-reading the text and providing continuous encouragement, is of course my partner, Barbara, to whom this book is dedicated.

Introduction

Much of this book is based on my experiences presenting "popular" science lectures and demonstrations to audiences around the UK. I have a keen interest in the "Public Understanding of Science" and in communicating my own enthusiasm for science to others. I firmly believe that a little appreciation of how science influences and shapes our lives is essential for any well educated person. However, many people feel alienated by the sciences. Often science can appear completely unapproachable and can be couched in veils of mystery and complexity.

Nevertheless, there is science all around us. For example, as I often point out, anybody following any recipe in the kitchen is in effect performing a scientific experiment. Cooks who learn from their experiences with recipes and manage to improve their skills are doing no less than scientists working in their laboratories.

Indeed, I apply the same methodologies when at work in my Physics laboratory, or at home in my kitchen. When preparing a dish for the first time, I follow a recipe. The recipe may come from a recipe book, or made up in my head, but there will always be a list of ingredients and a set of instructions to mix and cook them in the appropriate way. Once the dish is finished, then it is tested – by eating it! Then the results are analysed. Was the meal good? How could it be improved? And so on.

Then, when I next prepare the same dish, I make appropriate amendments to the recipe that I believe will provide the desired changes in the final dish. Once again it is tested and more improvements suggest themselves, on so on.

This process of continual revision of the recipe is just an adaptation of the experimental approach to science. However, as in all scientific experiments, a little understanding of the underlying theoretical side can help enormously in planning the modifications for the next experiment. In cookery, the greater the cook's understanding of the processes that occur to develop flavour, texture, etc.

the more likely he or she is to make changes that will quickly and effectively improve any recipe.

I hope that readers of this book will find they learn enough about the science that happens whenever they cook that they will be able not only to understand why things sometimes go wrong, but also be able to make sure failures are much less likely in their kitchens.

In Chapters 2 to 5 I have tried to provide an introduction to some of the science that lies behind the chemical and physical changes that occur in foods as we cook them as well as an appreciation of how we taste and savour our foods. Inevitably, those with a strong science background may find some of the explanations given here to be perhaps a little simplistic, while others who have forgotten most science they ever learnt may find parts quite hard-going. However, it is never possible to please all the people all the time. Instead, I have tried to set a level that should be sufficient to enable readers to gain some useful understanding, while not attempting to go into such detail that the important principles would be lost.

These opening chapters are really intended as a short primer and more importantly as a reference source so that interested readers can find out more whenever they think it may be interesting.

In the later chapters (6 to 13) I have tried to demonstrate the science behind a range of types of cooking by discussing specific illustrative recipes. In each of these chapters there is a short introduction where the main scientific points are briefly discussed (with appropriate reference to the earlier chapters as needed) followed by a few recipes to illustrate these principles in practice.

All the recipes presented in these chapters have been carefully written so that there is a good reason for every ingredient and for every instruction. I have tried to draw attention to the reasons why each instruction is there and to show what may go wrong if one is ignored.

I have also included with each main recipe some descriptions of things that could go wrong (usually from personal experience!) and some comments as to why these disasters may have happened as well as suggestions as to how to correct the problems.

Many of the chapters have separate panels of text describing interesting (to me) pieces of relevant science, etc. These are intended not only to lighten the text, but also as an aid to understanding of some of the science involved in the actual recipes.

After the main recipes in each chapter, I have tried to suggest some variations you may like to try – these are intended to stimulate you to invent your own recipes. I have found that once I understand how a particular type of recipe works it is very easy to adapt to produce different and interesting results.

Finally, at the end of most chapters I have described a few science experiments you might like to try at home. These experiments are intended to illustrate some of the scientific issues addressed in the chapters, while also in some cases also helping you to improve your cookery skills and in others just to have a bit of fun. However, a few of these experiments involve using heat and fire and have a small risk of danger attached. So please read very carefully right through the description of any experiment and make quite sure you understand it fully before moving on to try it out. Also always think to yourself – "what could go wrong here?" and "what would I do if that happened?" before actually trying any of the experiments.

Most of the experiments are suitable for people of all ages, but a few need adult supervision and all would benefit from the help of an adult. I hope that a few families will have some real fun and learn some interest in science from working their way through these experiments. If I can influence a few keen youngsters to take up a science career, I will be well pleased!

Sensuous Molecules – Molecular Gastronomy

Introduction

This chapter and the next two chapters are intended to provide a little background science that may help with the understanding of some of the later chapters. Many readers who already have some scientific knowledge can easily skip through most of this material. On the other hand, readers with little or no memory of the science they learnt at school may find some a bit hard going. Nevertheless, I hope most readers will find the material useful and perhaps, as they read the later chapters, recognise that a basic scientific knowledge can actually improve their cooking skills.

At first sight you may wonder whether there is any good reason why you should bother to work your way through all the minutiae of the science of atoms, molecules etc. and to what extent, if at all, it will improve your cooking skills. I firmly believe that the better the basic understanding you have of the way in which chemistry works, the better you will understand cookery in chemical terms and the more likely you understand how to improve your own cooking skills.

The usefulness of models based on a good understanding can be seen in many diverse circumstances. Nearly everyone drives a car these days and all drivers use models to understand how the controls work. At the simplest level, we all know that to go round a corner we need to turn the steering wheel. If we press down on the accelerator the car will go faster, or if we hit the brake pedal it will stop. Such simple models, that relate cause and effect, with no appreciation of why the two are related, will be adequate in most circumstances. Better models could, in some circumstances, be used to avoid accidents. Drivers who have a deeper appreciation of the steering system have a better model, that enables them to understand instinctively how to avoid skidding. Similarly a model that includes an understanding of what goes on inside the engine, can empower a driver to perform their own routine maintenance, or effect roadside repairs without having to rely on a mechanic.

In general, the quality of the picture, or model you have of any situation, depends on how deeply you understand the underlying principles. For example, if you play chess the simplest model consists of just knowing the rules of how to move the pieces. A better model involves some understanding that some pieces are more valuable than others; with a model of this type you will remember not to give away your queen lightly. A deeper understanding will come from an appreciation of strategy and knowledge of some basic openings and end games, at this level a player can challenge most (commercial) computer programmes. Deeper understanding of the way to analyse positions is needed before a player can reach a good competition standard.

The models most people use when cooking are based on their own experience and, to some extent, from the instructions they have read in recipe books. Thus we all know that boiling a small egg for 3 or 4 minutes will provide a soft, runny yolk; while boiling it for 6 minutes will produce a hard yolk. We use similar rules of thumb to work out how long we should cook the Sunday roast – typically we use some rule such as 15 minutes per pound (30 minutes per kilogram) plus 20 minutes extra. Such models will serve us well in many circumstances, but if we do not appreciate the underlying science behind them we will not realise their limitations. You may not often want to know how long to boil an egg at high altitude, in the Andes, or to prepare boiled duck eggs, or to roast an usually large (or small) joint, but you can be sure that the above rules of thumb will fail (as described in Chapter 4). With a reasonable level of understanding of the science, however, you will be able to work out how to cook the eggs and joint perfectly.

Good models can empower, but bad models can lead to disasters. In Zimbabwe, the people have a model of lightning based on their accurate observation that lightning normally strikes high points. In their culture they think of lightning as being a big bird that makes its nest in high places and often returns to the same spot. The problem with the model is that it has led people to place large metal spikes on top of their dwellings to prevent the bird from building its nest on their roofs. As a consequence of all these spikes Zimbabwe has the highest death rate from lightning strikes anywhere in the world. Of course, better educated people know better and understand that lightning is attracted to metal spikes on top of high buildings. ...But do you have a television aerial on top of your house?

You can think of a recipe as a very simple model, most cooks use recipes as a rough guide, they interpret the instructions to suit their own equipment and modify the ingredients according to personal preferences and tastes as well as to the available materials. Suppose you want to make a Brie soufflé; you might find a recipe for a cheese soufflé from a cookery book, or, you might try to modify a recipe for a sweet soufflé. In either case you will find it quite difficult to obtain a perfect result since most recipes will produce soufflés that are liable to collapse rather easily (as described in Chapter 12). However, if you understand the underlying chemistry, you will understand the importance of preventing the fat in the cheese from destroying the egg foam, so you will either find a recipe that has a good chance of working, or you will think up, for yourself, a way of encapsulating the cheese in some inert medium, such as a starch based sauce.

Chemistry is all about the way in which atoms join together to make molecules. There are only about 100 different elements in the whole Universe, the smallest "indivisible" unit of each element being an atom. In practice, when considering food and cooking we need only be concerned with a few of these elements; most of what we cook and eat is made up of carbon, oxygen and hydrogen with a smattering of nitrogen and traces of sodium, sulphur, potassium and a few others.

It is useful to develop a good understanding of how chemistry works; what happens in chemical reactions and how to classify molecules into different types. It is best to start with a model of atoms and then to consider the ways in which atoms join together through chemical bonds to form simple molecules such as water. From such a well founded but simple start it is relatively straightforward to go on to the important types of molecules found in food – sugars, fats, proteins, etc. and then to begin to understand how these different types of molecule react with each other when you are cooking even the simplest dishes.

It is often said that we are what we eat, and in large measure this is quite true. When we eat food, our bodies break down the complex molecules in the food into simpler, smaller molecules. We then use these simple molecules as building blocks to manufacture the more complex molecules we need to live. We need to make skin, bones, muscles, blood, etc. all the different components that make up our bodies. Most of these we can synthesise from a fairly wide variety of starting foods, but there are some essential substances that we just can't make at all. These are the vitamins, which we need to eat in sufficient quantity to live at all. Other classes of molecules that we need to make to sustain life include the proteins, the sugars and fats.

Atoms and Molecules

Much of what happens in cookery is best described as chemistry. The processes by which different atoms (or molecules) are brought together and made to form new molecules are generally called "chemical reactions". The development of 'meaty' flavours on heating and browning is caused by some complex chemistry called 'Maillard reactions'. The fact that boiled eggs set hard is due to chemical reactions between the proteins in the egg. Food sticks to pans during cooking because proteins react chemically with metals at high temperatures.

In this section we will first review the concepts of atoms and molecules before moving on to look in greater detail at some of the more important groups of molecules found in food.

The ancient Greeks worked out that everything was made up from small building blocks. Ideas about the size and composition of these building blocks have changed greatly over the intervening years, particularly in the last century. The present level of understanding is that all matter is made up from very small pieces known as molecules. Molecules are made up from smaller particles called atoms that are joined to each other. Atoms in turn are made up from even smaller particles, protons, neutrons, electrons etc. (collectively termed sub-atomic particles). Today, physicists think that the smallest particles in nature are quarks; every sub-atomic particle being made up from three quarks.

Before discussing atoms and molecules further, you should have some concept of just how incredibly small they are. Suppose you have a glass of wine and ask how many atoms are there in the wine? The answer is, roughly, 10,000,000,000,000,000,000,000,000 (usually written as 10^{25}) which is so large that nobody can really comprehend how big it is. One way to try to conceive the magnitude of the number of atoms in a glass of wine is to consider how much space would be taken up by the same number of grains of salt. Suppose you took 10^{25} grains of salt and spread them on the ground 1 metre deep; the resulting salt pile would spread over the entire surface of the earth; land and sea!

The models of the structure of the atom have changed a good deal during this century, as techniques to probe into the microscopic world they occupy have improved. As our knowledge of the behaviour of sub-atomic particles improves so the models we use to understand them seem to become crazier and wackier. These days we imagine sub-atomic particles to inhabit a world where probability rules. The only fact we can be certain of is that our model is wrong. Probably.

Before moving on to describe a simple (and for our purposes adequate) model of an atom, you should be aware of three of the sub-atomic particles that go towards making up atoms. These are the electron, the proton and the neutron. Protons and electrons are electrically charged. You are familiar with electric charge – or at least with the movement of charge. Lightning strikes when the charge in a cloud is released to the earth. You get a sudden shock when a static charge that has built up when you walk over some carpets is released when you hold a metal object that is itself earthed, such as a car door, etc. The amount of charge on an electron is exactly the same as, but of the opposite "sign" to, that on

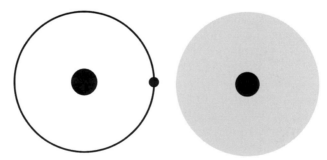

Figure 2.1. Left: a simple, planetary model of an atom, with the positively charged nucleus at the centre and the negatively charged electrons in fixed orbits a long distance away and, right: a better model where the electrons are in cloud-like shells centred on the nucleus

a proton (i.e. electrons are negatively charged while protons are positively charged). Charges of similar "sign" repel one another, while charges of opposite "sign" attract each other. The neutrons are electrically neutral. Although all these particles are vanishingly small (even when compared to an atom) the proton and neutron are, relatively, very much larger than an electron.

Fortunately, there is no need here to understand the structure of atoms in any greater detail. A useful picture of atoms that is at least consistent with the current models, is one consisting of a nucleus (made up of positively charged protons and neutral neutrons) with some electrons whizzing around in orbits a long way away. Since atoms are electrically neutral, the number of electrons is equal to the number of protons in the nucleus. Different atoms have different numbers of protons. For example Hydrogen has one proton, Carbon has 6, Oxygen has 8 and Uranium has 92.

A rather outdated, but nevertheless useful model is to consider the electrons to be in fixed orbits rather like the planets around the sun. If you were to imagine the nucleus to be the size of the sun, then the electrons would be less than the size of a pea and would be about as far away as Jupiter! A better picture is to think of the electrons not as separate particles but rather as distributed into thin clouds spread around the nucleus in thin shells.

Each of these cloudy shells likes to have a particular number of electrons in it. Atoms with just one or two electrons in a shell will share these spare electrons with another atom with an almost full shell. This is the way in which molecules are built up with chemical (or covalent) bonds joining together the atoms as they share their electrons. A simple way to think of these bonds is to consider that the atoms with excess electrons have 'hooks' (the number of 'hooks' being equal to the number of excess electrons in the outer shell) while atoms with almost full shells have 'eyes' that can link with the 'hooks'. Thus Hydrogen has one 'hook' and Oxygen has two 'eyes' so two hydrogen atoms will combine with one oxygen atom to make a molecule of two Hydrogen atoms (written H_2) and one Oxygen atom (O) making H_2O or water. Similarly, Carbon has four 'hooks' so it combines with two Oxygen atoms to make CO_2, or carbon dioxide.

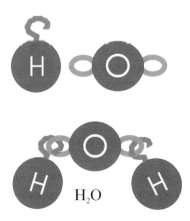

H_2O

Figure 2.2. A sketch to illustrate how water molecules are formed by combining two Hydrogen atoms and one Oxygen atom

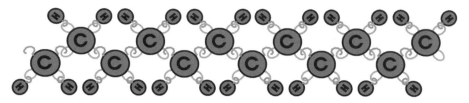

Figure 2.3. A simple model of part of a polythene molecule – the large circles represent carbon atoms and the small circles hydrogen

When not many atoms (up to maybe as many as 50 to 100) are involved in making a molecule, we tend to say it is a small molecule. In food it is these small molecules that carry the smells and flavour, as we shall see in Chapter 3. There are two particularly important classes of small molecules involved in food and cooking, fats and sugars. These molecules are described in a little detail later.

Two types of polymers are of particular importance in cookery. Proteins and starches are long molecules made up by joining together many separate repeating units. We call such molecules long chain polymers. Proteins are made up from units called amino acids while starches are made up by stringing together sugar molecules. The structure of proteins and starches are addressed later in this chapter.

Fats and oils

Many people seem to think there is some great difference between a "fat" and an "oil". There is no real difference. We just happen to use the term "oil" for those materials that are liquid at room temperature, and call those that are solid at room temperature "fats". The distinction is thus rather arbitrary. Since we nearly always melt fats when we cook with them they should be called "oils" in use.

All fats and oils produced by plants and animals are used to store energy. The molecular structure of oils and fats is very similar to that of the fuel you use to power your car. Just as with the petrol in a car engine, the energy is released when the fat is "burnt". Burning, in this context, really means "oxidation"; the reaction of the fat with oxygen. This reaction generates a lot of heat; in a car engine, the reaction is very fast, so that a lot of heat is generated in a small space in a short time. In fact so much heat is produced in such a very short instant of time, that there is an explosion inside the cylinder of the engine, which drives down the piston, which turns the crankshaft, which drives the car. The difference is that in our bodies we manage to carry out the reaction more slowly in a controlled fashion so that the heat is not so concentrated, but nevertheless a similar amount of energy is liberated.

Essentially fats and oils consist of short strings of carbon atoms with hydrogen atoms attached to them. Most fats we encounter in cooking have three

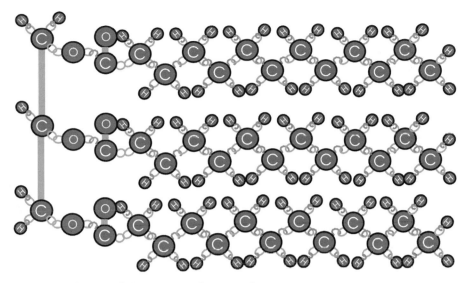

Figure 2.4. A model of the structure of a typical fat molecule

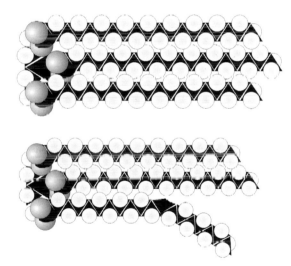

Figure 2.5. Top a saturated fat and bottom a mono-unsaturated fat. Note the kink in the unsaturated fat

strings all joined together at one end. Typically there are about 10 to 20 carbon atoms in each string.

One of the most important ways in which fats are distinguished from one another is the "degree of saturation". A saturated fat has the maximum possible number of hydrogen atoms attached to the carbon atoms. That is to say that each carbon atom in the chain is attached to two hydrogen atoms and two carbon atoms (we say the carbon atoms are joined by "single" bonds). When fats are made in this way they can pack together very easily. The chains lie down in almost straight lines (actually they form "planar zig-zag" structures that can pack simply side by side). In unsaturated fats two or more of the carbon atoms are joined together by a "double bond" and have only one hydrogen attached. In mono-unsaturated fats just two carbons are joined by such a double bond, in poly-unsaturated fats several pairs of carbons are each joined by double bonds. When a double bond is introduced along a chain of carbon atoms the regular zig-zag shape is interrupted and the chain becomes kinked. It is very much more difficult to pack such kinked chains neatly side by side than it is to pack the well ordered planar zig-zag saturated chains.

It is this difference in packing that provides the reason for whether a fat is solid, or liquid. The saturated fats that can pack easily together form solids more readily and have higher melting points. They also store more energy. So in mammals, where the body temperature is maintained around 35 °C, saturated fats will generally be liquids, and hence will be readily accessible to use as and when needed. So mammals tend to produce the more efficient saturated fats. In fish and plants where the temperature can be much lower, saturated fats would tend to solidify and become difficult to use, and may even form constrictions in the circulatory systems leading to disease, etc. So they tend to use the less efficient unsaturated fats for energy storage.

The main reason why we regard saturated fats as being less healthy than unsaturated fats lies in their higher melting temperature. If we should get a build up of saturated fats in our arteries, and the particular fat should have a melting point close to, or even above, body temperature, there is a real danger that the fat may solidify in an artery and hence cut of the flow of blood – leading to high blood pressure or even a stroke.

Another important difference between the saturated and unsaturated fats is that the double bonds in unsaturated fats are easier sites for oxidation. In fact, even at quite low temperatures, atmospheric oxygen can react with unsaturated fats. When such oxidation occurs we recognise it as the fat "going rancid". The off flavours of rancidity are actually caused by oxidation of the fat. Saturated fats, such as beef dripping, are very stable at room temperature. Oxygen cannot easily penetrate into their solid structure and the fact that there are no double bonds makes the rate of oxidation at the surface very slow. In contrast, if a vegetable oil is left open at room temperature it quickly oxidises and becomes rancid.

Butter, is a rather special case; although the fat part of butter is largely saturated, it nevertheless can easily become rancid. This happens through a different process (hydrolysis), which is promoted by the water droplets in the butter. "Clarified butter" from which all the water has been removed by gentle heating until it stops bubbling, is much less prone to rancidity.

Sugars

To most people sugar is the sweet, white crystalline material that comes in packets from the supermarket. To chemists, sugars are a well defined class of chemicals, of which "sucrose" (the sweet, white crystalline substance in the packets) is just one member of this family of molecules. For the purpose of this discussion there is no need to enter into the formal definition of a "sugar", but there is a need to realise that there are several different sweet, white, crystalline substances all of which are collectively referred to as sugars.

Table sugar, once refined from sugar cane, or sugar beet, is 98% pure sucrose. The sugar in honey is almost entirely fructose, the sugar produced in milk is mostly lactose. So what are the differences between all these sugars? And do they matter?

The important food sugars are all made up of rings of 4 or 5 carbon atoms and one oxygen atom, with 1 or 2 more carbon atoms attached on the side of the ring. Some, such as glucose, are made up of just a single ring, while most such as sucrose consist of two rings joined together. The collective name for the sugar rings is "saccharides" so the single ring sugars are called mono saccharides, and the two ring sugars are called di-saccharides. If several sugar rings are joined together the resulting molecule is usually called an oligo-saccharide.

Sugars, like fats, are produced by living organisms to store energy. In general, because sugars already contain some oxygen, the efficiency of energy storage is lower, but the very presence of the oxygen makes it easier to burn sugars so the energy can be released more readily. The single ring sugars will generally release more energy when burnt, than the multiple ring sugars. However, in the body, we need to control the oxidation reaction (often called the combustion) of the sugars so we use special molecules (enzymes) that perform the actual chemistry – you can think of enzymes as miniature test-tubes. The enzyme needed for a particular chemical reaction will only drive that reaction and no other, so there are different enzymes for the reduction of the different sugars.

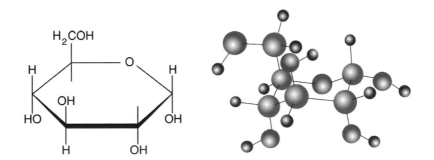

Figure 2.6. Two different types of model of a glucose molecule – on the left is a model illustrating the chemical bonds and on the right is a model where the atoms are drawn as small balls

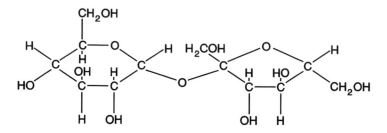

Figure 2.7. A diagram of a sucrose molecule. Sucrose is made from two sugar rings joined together. It is called a disaccharide molecule

Accordingly, different animals and plants tend to produce different sugars to store energy, depending on the enzymes they have evolved over the millennia. Most plants produce sucrose, while most mammals tend to produce the disaccharide, lactose. Man is able to digest more or less any mono- or disaccharide, although we need to convert the sugars into digestible forms before we can use them. However, we do not possess any enzymes that are capable of breaking down larger sugars, such as raffinose, etc. These 3, 4 and 5 ring sugars are made by many plants especially as part of the energy storage system in seeds and beans. If we ingest these sugars, we can't break them down in the intestines, rather they travel down into the colon where various bacteria digest them – and in the process generate copious amounts of carbon dioxide gas, with results that can be both unpleasant and anti-social.

Polysaccharides and Starches

Some of the most important biological molecules are made by joining very many sugar molecules together to form long strings – molecules of this type are called "polysaccharides" and belong to the general class of molecules known as carbohydrates (molecules made up from carbon, oxygen and hydrogen atoms). Since there are several different sugars (or mono-saccharides) that can be used as the building blocks, and since these can be put together in any order, there is an infinite possible list of polysaccharides. From this list, nature has made a very large selection, with an enormous range of properties and uses. Three of the commonest of these polysaccharides are cellulose, amylose and amylopectin (amylose and amylopectin are the main components of starch). These three polymers have very different properties and behave very differently in cookery and in the digestive system. Cellulose is the constituent of plant cell walls that gives them stiffness and strength. Since the plants do not wish to break down their own strength, they do not posses enzymes capable of digesting cellulose. In fact cellulose is, biologically speaking, a rather inert material, very few organisms can break it down and digest it and it is more or less insoluble. In contrast, starch forms a staple food of many plants and animals; all plants, and many ani-

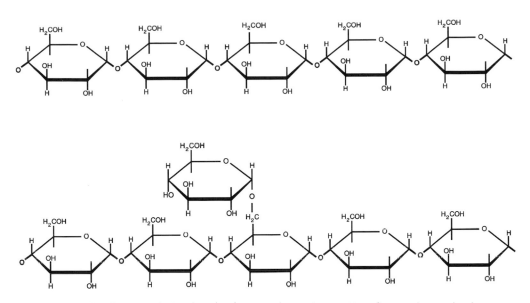

Figure 2.8. Diagrams of starch molecules — at the top is a section of an amylose molecule — the linear starch molecule and below it is a section of an amylopectin molecule — the branched molecule

mals possess enzymes that allow them to digest starch. Starch can be swollen and in some cases dissolved in water which greatly assists the digestive processes.

However, despite these widespread differences between cellulose and starch, they have exactly the same chemical formula. Both are made up rings of glucose joined together end to end. The difference is in the geometry of the way they are joined. The cellulose molecules are joined in such a way that they make stiff molecules that pack tightly together and are held in place by powerful internal hydrogen bonds. So tightly do the cellulose molecules pack together that it is very difficult to prise separate molecules apart which is a necessary first step in their breakdown. So cellulose remains a very stable material. On the other hand, the geometry of the joins between the sugar rings in the starch molecules leads to a more open helical structure and fewer internal bonds, so that while amylose and amylopectin molecules do pack together, they do so in a looser and weaker fashion and are easily separated ready to be broken down by the specific enzymes that all plants and most animals possess for the purpose.

As I have already noted in passing, starch consists of two molecules, amylose and amylopectin. Both these molecules are made up from glucose rings joined together. The difference lies in that in amylose, the sugar rings are always joined end to end making up a linear molecule; while in amylopectin, some of the sugar rings join to three, rather than two others which leads to a branched molecule with a structure more like a bottle brush, than a length of string.

Starch granules

Starch is formed by many plants in small granules – a typical granule may be a few thousandths of a millimetre across. Within a granule the plant will lay down successive rings, with a higher, or lower, proportion of amylopectin. In the rings with a lower amylopectin content the molecules are packed close together in a well ordered form, making these parts of the granules more resistant to attack from enzymes. Biologists often refer to the more ordered layers as "crystalline" layers. Of course, the granules are not purely amylopectin and amylose, the plants also incorporate some proteins as they make the granules. Importantly, different plants (and different varieties of the same plant) incorporate widely differing amounts of protein in their starch granules.

From a cooking point of view the amount of protein and where it is located in the starch granules is crucial. When cold water is added to starch granules it is absorbed by the proteins, but hardly penetrates the amylose and amylopectin. Thus, starch granules with a high protein content will absorb a lot of moisture at room temperature, while those with low protein contents absorb but little water.

The absorption of moisture is very important for two reasons. First, when there is some moisture around, then bacteria can become active and begin to digest the starch. Secondly, once the proteins around the outside of a starch granule have absorbed some water, then they become "sticky" and will bind granules together into large lumps. Once a large "lump" has been formed, those granules near the centre find themselves well sheltered from the outside, so that they are unlikely to be attacked by any external bacteria, and more importantly are unlikely to be further swollen by any additional water.

We come across starch granules in all forms of flour – in fact flour is really just a collection of starch granules. We keep flour dry when it is stored both to keep it from deteriorating through bacterial action, and to prevent it from forming lumps as the proteins, if wetted, will stick together.

Gluten

There are two main proteins in the starch granules from wheat flour: gliaden and glutenin; these two proteins can combine together with some water (about twice as much water as protein) to form a complex known as gluten. To form gluten, the gliaden and glutenin protein molecules need to be stretched out side by side and surrounded by some water. Once the proteins are stretched out they begin to interact and form new bonds between each other. The resulting gluten is a complex network system of mutually interacting molecules. This network is highly elastic and very tough. Gluten formation is of the utmost importance in bread, cake and pastry making, as we shall see in later chapters. Suffice it to say here that to make good bread you knead the dough to produce gluten, while to make good pastry, you treat the dough very gently to avoid the formation of any gluten.

Swelling on heating

While cold water will not greatly affect the amylose or amylopectin in a starch granule, hot water certainly will. When starch granules are heated the crystalline layers start to melt as the temperature exceeds 60 °C. The actual melting temperature depends on the relative amounts of amylopectin and amylose and on how well the amylose molecules pack together to form small crystals inside the granules. As soon as the crystals are melted, the separate amylose and amylopectin molecules become much less well packed together and tend to move apart. This disordering and opening up of the structure of the granules allows water to penetrate. The linear amylose molecules are quite soluble in the water and the branched amylopectin less easily dissolved. As the molecules overlap with one another to a significant degree, they do not fully dissolve in the water, but rather form a soft gel. Starch granules can absorb an enormous amount of water without losing their integrity – this is one of the reasons why they are such good thickening agents. For example, potato starch granules can swell up

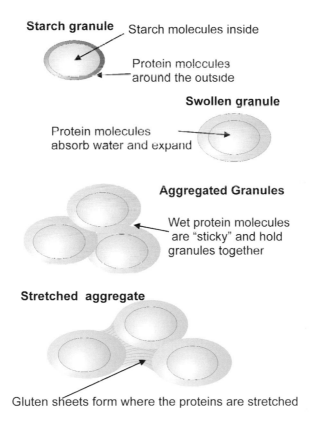

Figure 2.9. A series of diagrams illustrating how gluten is formed by stretching of starch proteins

to 100 times their original volume without bursting. Once a starch granule bursts, the separate amylose molecules can become dissolved in the surrounding water.

The swelling process is not fully reversible, if a swollen starch granule is dried, then the original order in crystalline layers of amylose is not restored. The molecules do form an ordered structure, but one of a different crystal form. In this new form the amylose molecules adopt a new helical form in which they are intrinsically bound up with some water molecules. Such a combination of a starch molecule with water molecules is often called a "complex". The water in these amylose complexes remains bound up in the crystals, giving the impression of the system "drying-up". This is the process we normally call "staling".

It is important to remember that, if the starch granules are allowed to stick to one another when cold water is added, they will form lumps. If a mixture with such lumps is subsequently heated, then water will not penetrate into the centre of these lumps and the result will be a very uneven texture. In fact this is the recipe for lumpy gravy! So to avoid lumpy sauces and gravies, it is essential to prevent any granules, whose proteins may have absorbed some water and hence become sticky, from coming together until the temperature is high enough that water can penetrate and swell the starch. Several methods can be used, as you will see in Chapter 9. Essentially these all involve separating and suspending the granules either in some fat, or for granules with only a small amount of protein in cold water, before heating.

Proteins

There are many molecules that are essential for life, foremost among these are proteins. To make proteins we need to eat proteins. This is why all diets call for an element of protein. Although you may think of proteins as just one class of foods referred to on the back of packaging where the nutritional value of food is listed, they are much more than just foodstuffs.

Proteins are a special class of polymers made up by joining together amino acids – each amino acid is made up of about 20 atoms. There are more than 20 different amino acids and most of them occur in most proteins. This diversity of building blocks means that there is an enormous number of possible molecules that can be made. In the UK National Lottery players are asked to select 6 numbers from a pool of 49, there are almost 14,000,000 ways of making a selection – which is why winning is so unlikely! When making proteins you can pick, in sequence, any number from 50 to 10,000 of these amino acids (you can pick the same amino acid as many times as you wish). There are, therefore, far more possibilities than there are on the lottery!

The diversity of possible protein structures is at the heart of their biological importance. It is possible to select from the almost infinite range of possible protein molecules ones that have particular shapes and perform specific tasks. It is

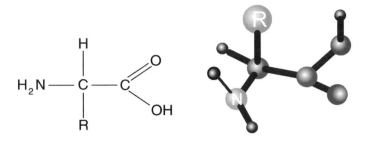

Figure 2.10. Two diagrams showing the structure of an amino acid. The group marked "R" takes different forms in the different amino acids

the shape of the proteins that gives them their particular biological function. For example, the protein haemoglobin is designed to carry oxygen around in the blood. Haemoglobin has a shape that leaves a 'hole' into which oxygen atoms fit perfectly; when the haemoglobin reaches a muscle that wants some oxygen the muscle sends a chemical signal that causes the haemoglobin to change its shape and the oxygen pops out. At the same time (and as a consequence of the shape change) the haemoglobin changes colour from red to purple.

Proteins are essential molecules for all life, their shapes allow them to control many diverse biological processes. The shape of a protein molecule is determined both by the sequence of amino acids along its length and by internal 'bonds' between different amino acids; a different sequence of amino acids will produce a protein with a different shape.

There are several types of internal bonds that can form links between the amino acids in proteins; these are given fancy names such as 'disulphide bridge' or 'hydrogen bond'. There is no need to know much about these internal bonds, other than that they exist, they help define the shape of the proteins and that they can be broken by various methods (which will be discussed later). Indeed, despite the scientific sounding terms there is much that is not known about the way in which these internal bonds operate. A good definition of a hydrogen bond (paraphrased from Prof. John Polanyi who won the Nobel Prize for Chemistry in 1986) is: "A hydrogen bond is chemist-speak for 'it makes atoms stick together but we don't know why'".

If the internal 'bonds' of the proteins are broken then their shape changes from that in the 'natural' state; this process is called denaturation. Many proteins are rather tightly coiled up (globular proteins) so when the internal bonds are released, and they become denatured, they expand outwards. You can imagine a protein in the natural state as a tightly wound ball of wool; all proteins of the same type would be represented by identical balls of wool. If a kitten were allowed to play with the balls of wool they would quickly become random tangles, this is what happens to proteins when they are denatured.

The most usual cause of denaturation in cooking is heat. All molecules vibrate all the time. The amplitude of these vibrations increases as the temperature is

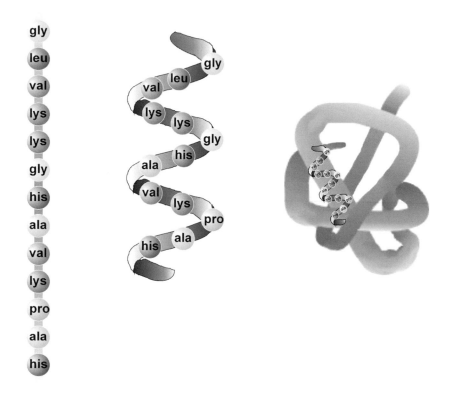

Figure 2.11. Diagrams illustrating protein structure. On the left is a simple representation of a part of a protein molecule made up from a sequence of amino acids (each amino acid is given a three letter code – e.g. "gly" for glycine, etc.). In the middle is a more realistic representation of a section of the protein that has formed a helix. At the right the whole protein is sketched just showing its overall shape with a small helical section highlighted

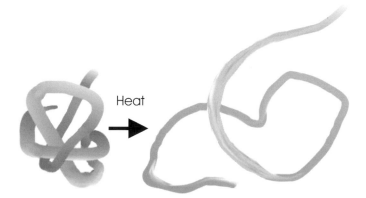

Heat

Figure 2.12. When a protein molecule, drawn here as a line, is heated the internal bonds are released and it changes its shape. The process is called denaturing

Sensuous Molecules – Molecular Gastronomy

increased. In proteins if the vibrations are strong enough then the molecule can literally shake itself free of its internal bonds. People use this property of proteins to fight infection; viruses are complex molecules which can be very sensitive to heating. When we are ill, our immune system raises our body temperature so as to denature the proteins coating the viruses, preferably without getting to so high a temperature that the proteins in our bodies are denatured (in which case we would die!).

Most proteins are denatured at temperatures around 40°C. When proteins are heated to higher temperatures, they start to undergo chemical reactions that can cause them to break up or to join together into even larger molecules. These chemical reactions are at the heart of cookery. When we cook an egg the egg proteins denature once the temperature is above 40°C, and they start to react together to 'cook' the egg once the temperature is above about 75°C. Once these reactions start the egg changes from a liquid solution of proteins into a solid mass. You could imagine this process in terms of our kitten with the balls of wool. Suppose that the floor is covered with separate balls of wool, the kitten plays with all the balls and they get denatured into random tangles on the floor. Now as we heat further we introduce more kittens and all the tangles get knotted together into a solid mass (as the new chemical bonds are formed between different denatured protein molecules).

Another important way to denature proteins is by stretching the molecules. When a solution of protein molecules flows the molecules can become stretched. Stretching happens when the liquid is accelerating, all the liquid becomes stretched out and, provided the flow is fast enough, the proteins dissolved in it can be extended. In a static solution the proteins are in the natural state, they are tight coils; in the flow the proteins can stretch out and become long strings. The flow fields produced between a pair of counter rotating beaters – as in an egg whisk – and between a balloon whisk and the sides of a bowl both create ideal conditions for denaturing proteins.

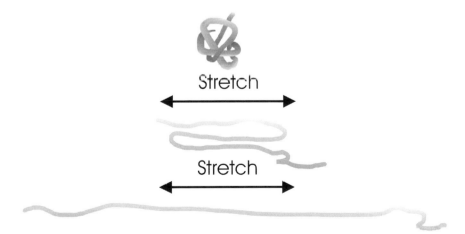

Figure 2.13. A sketch showing how a protein can be denatured by stretching

Collagen, Gelatin and 'Gels'

The protein, collagen, deserves a special mention. Collagen is a stiff fibrous protein that is abundant in all mammals. Collagen is a major component of skin and of the sinews that connect muscles to bones, as well as forming coatings around bundles of muscle fibres.

Collagen is not a single molecule, but consists of three separate molecules twisted around each other in a rope like structure. It is this "triple helical" arrangement that provides collagen with its stiffness and makes it so useful as a structural building block in mammalian tissue. However, this same stiffness and toughness makes collagen almost inedible. Before we can digest collagen we need to break it down into its three separate strands – these are flexible protein molecules that are readily digestible.

Figure 2.14. A sketch of the collagen triple helical structure

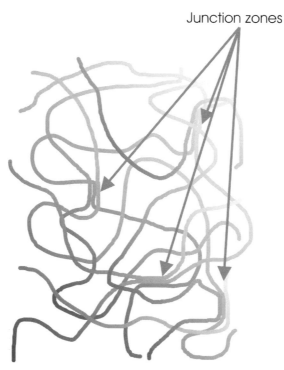

Junction zones

Figure 2.15. A sketch showing how a network can be built up from a series of molecules that interact in a few "junction zones"

When collagen is heated above about 70°C the separate strands of the triple helix unwind, denaturing the collagen into its separate molecules – known as gelatin. The denatured, gelatin, proteins cannot reform into a triple helix when cooled. Instead, the separate strands interact with one another forming many links with other molecules and so build up a large network.

Between the junction zones of the networks, the gelatin molecules remain "dissolved" in the surrounding water – more strictly we may say that water is bound to the gelatin molecules. Since there is a molecular pathway right through the whole network, the gelatin water system behaves like a solid, rather than a liquid; even though it may contain as much as 90% water. We call such systems "gels" or in the kitchen jellies.

The gels formed by gelatin are "thermo-reversible" gels. That is to say that if we increase the temperature, the links between the separate molecules weaken and break (as the temperature exceeds about 30°C) so "melting" the gel. Similarly when we cool the dissolved gelatin below about 15°C, the separate molecules start to interact again and the gel reforms.

Gels are also formed from other molecules in cooking. For example when egg whites are heated the proteins denature and then form new crosslinks to build up a gel – which we recognise as cooked egg white. In this case, the gel is permanent. The links between the egg white proteins are formed by irreversible chemical reactions.

Soaps, Bubbles and Foams

Just like people and animals, some molecules like water and others hate it. For example oils and fats, like small boys and cats, are repelled by water. Scientists as usual make up complicated words to describe such simple concepts. Molecules that like water are called hydrophilic, while those that dislike water are termed hydrophobic. It is possible to make special molecules that have one hydrophobic end joined by a flexible chain, to a hydrophilic end. These special molecules are very important in nature, since they are the basic building block for the walls of all cells. Molecules of this type are called lipids when they are made in nature, or soaps or detergents when made by man.

While there are some technical differences between detergents and soaps they both work in much the same way, they have one (hydrophilic) end that wants to be surrounded by water and one (hydrophobic) end that will stop at nothing to get out of the water. Thus, in a washing machine, the hydrophobic ends will coat any dirt (oils fats, etc.) while the hydrophilic ends remain in the water. The result is that the dirt is broken up into small particles each of which is surrounded by a layer of detergent molecules with their hydrophobic ends firmly stuck to the oily dirt surface. So the dirt particles are encapsulated in small droplets that are stabilised by the hydrophilic ends and suspended in the water. In the diagram you can see how the hydrophobic ends of the soap molecules (rounded ends in the diagram) coat the dirt particle sticking to its

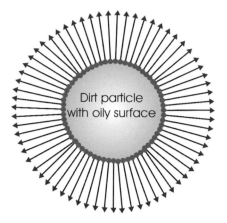

Figure 2.16. A sketch of a dirt particle surrounded by soap molecules. The "round" ends of the soap molecules near the dirt particle are hydrophobic (they dislike water). The triangular ends are hydrophilic (they love water) so the as the water sees only the hydrophilic ends of the soap molecules the dirt particle can "dissolve" in the water

oily surface, while the hydrophilic ends (triangular in the diagram) remain happily in the water and so stabilise the dirt in suspension in the water.

Everybody is familiar with the fact that when soap is added to water it will readily make bubbles. In clean water there is nowhere for the hydrophobic ends of the soap molecules to go. So the hydrophobic ends get together and make the soap molecules form thin films (called membranes) where the hydrophobic groups are all together and keep the water out.

These membranes can be quite strong themselves; if any air is introduced, the membranes can become curved and make bubbles. To make a membrane bend some force is required, the more curved a membrane becomes the more force is needed. In a bubble the force that bends the membrane comes from the pressure

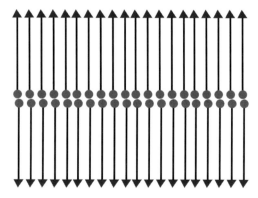

Figure 2.17. In clean water soaps form bi-layers with their hydrophobic ends in the middle of the bi-layers

Figure 2.18. When a soap bi-layer bends around on itself to form a sphere it can support an air bubble inside

inside the bubble, so smaller bubbles have a higher pressure inside them than large ones, because they have a smaller diameter and are more 'curved'. A foam is simply a collection of bubbles all stuck together. Smaller bubbles, with their higher pressure give stiffer, and stronger, foams.

Proteins can be made to behave in much the same way as soaps. Some of the amino acids (the groups of atoms which make up the proteins) are hydrophilic (they like to be surrounded by water) and others are hydrophobic (they will avoid water). In the natural state proteins are arranged so that the hydrophobic groups lie inside the tight coils and are not exposed to water, while the hydrophilic groups are on the outside and ensure that the proteins are well dissolved in the surrounding water.

Once the proteins are denatured the hydrophobic groups become exposed and they will try to get out of the water. One route to get away from the water is to go to the surface and emerge in the air. However, there is usually only a limited area of such surface present so the exposed hydrophobic groups will coat the surfaces of any fat or oil drops present in the liquid. Then if the liquid is stirred and the oil drops are broken up so increasing the available surface, more hydrophobic groups can move in and help to stabilise the new interfaces. Similarly, if air bubbles are made in the liquid, by for example, whisking, hydrophobic groups from denatured proteins will tend to form a thin membrane that stabilises the bubbles.

Bubbles and Foams:
Some experiments to try for yourself

A soap film can be very strong. You can easily make soap films by dipping a loop of wire, or even a loop made by bringing your forefinger tip to touch the tip of

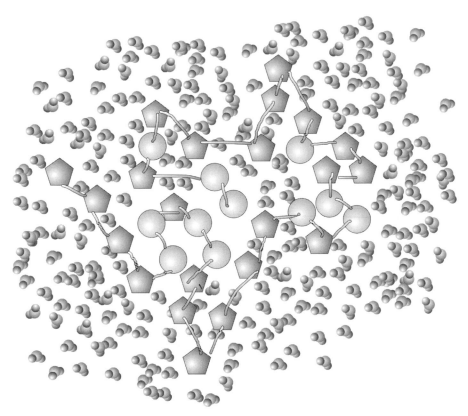

Figure 2.19 a. A simple sketch of a protein molecule in water. The hydrophobic patches along the protein are drawn as round balls, while the hydrophilic regions are represented by pentagons. Note how the hydrophobic parts lie in the middle of the coiled up protein and are thus shielded from the surrounding water (represented by the groups of three small balls)

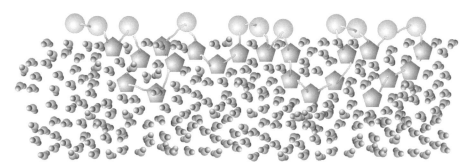

Figure 2.19 b. When the protein is denatured and is at the surface of the water, then it will rearrange itself so that the hydrophobic regions lie in the air and not the water. In this way a denatured protein can act in much the same way as a soap or detergent

your thumb, into a bowl of soapy water and then lifting it out carefully. A soap film will have formed inside the loop, if you blow gently on this film it will bend outwards forming a bowl shape. If you keep on blowing the film may bow out so much that it breaks away from the edge of the loop and forms a separate bubble. A bubble is simply a curved membrane formed from lots of soap molecules that form a sphere enclosing some air. The more curved the membrane, the higher the pressure of the air inside the bubble. So as the water in a membrane starts to evaporate, the membrane becomes a little less stiff and the bubble will expand under the pressure of air inside – usually this is a run away process and the bubble bursts. Occasionally, however, the bubble may slowly expand before your eyes – try watching a lot of bubbles and see if you can observe this phenomenon.

A foam is simply a raft of many bubbles joined together. Generally we think of a foam as having so many bubbles that we can no longer distinguish any separate bubbles. In a foam the many soap membranes join together and lose their curvature. Try making a wire frame in the shape of a cube and dipping it into a bowl of soapy water. Observe the shape of the soap membranes that appear on the frame when you lift it out – try blowing on the membranes to see how their shapes change. Notice that now several bubbles are joined together the membranes are no longer spherical, but tend to be made up of lots of flat segments.

Now remember that the hydrophobic (water hating groups) on the ends of the soap molecules really hate water, but they quite like oils and fats, so try seeing what happens to soap films and bubbles when they come into contact with some oil, or an oily surface. You will find that oils destroy soap films, burst bubbles and collapse foams. This is an important lesson as it shows how and why soufflés and some cakes tend to collapse. Understand the stability of foams and you will never fail to make perfect soufflés.

3

Taste and Flavour

Introduction

When we eat food, we often comment on its taste and flavour, but most of us don't really trouble to define what we mean by taste and flavour, nor do we stop to wonder whether there are different meanings to taste and flavour. Instead we simply enjoy the pleasures of eating.

Enjoying good food is an experience of all the senses. We see the food on the plate and anticipate a great meal. As the food is bought to us we can smell some of the aromas which whets our appetite further. When we put the food in our mouths we first taste it on the tongue and then experience further flavour sensations from the olfactory sensors in our noses as we breathe while chewing. The sounds the food makes as we bite into it also affect our enjoyment. The sharp crack of chocolate or pork crackling really add to the whole sensation of eating.

In this chapter I begin by describing the ways in which we taste and smell our food and describe a little about how flavours are released from foods as we eat them. In the second part of the chapter some of the chemistry involved in creating new flavours through cooking will be discussed – especially the very important class of "browning reactions" known as the Maillard reactions.

The Sense of Taste

We taste things with our tongues. We can detect five basic tastes – four are very familiar: sweet, sour, bitter and salt. The fifth, while familiar in the East is less well known in Western cuisine – it is called Umami and is the taste of monosodium glutamate, MSG. MSG is used widely in Eastern cooking and that is probably why it is recognised as a separate taste sensation more readily by those familiar with that cuisine. However, many common western foods contain large amounts of MSG, notably tomatoes and parmesan cheese.

There are many different molecules that trigger off each of the taste sensations. Taste buds that are receptors for salty taste react to many compounds besides common table salt (sodium chloride). Most sodium salts (that is most simple molecules that contain sodium) and most chlorides (that is most simple molecules that contain chlorine) will taste salty to greater or lesser extents.

Bitterness comes mostly from alkaloids (two common examples are quinine and caffeine). However, many alkaloids are poisonous, which may explain our general aversion to bitter flavours. Sourness comes from acids in our food – all acids provide a sour sensation, while sweetness comes from many other sources besides sugars.

There are many thousands of taste buds in the surface of the human tongue – exactly how these work and just what they respond to is still not fully understood. Indeed, researchers argue about how many different types of sensor there are for each taste.

The taste buds react to chemicals in food that manage to bind in some way to the surfaces of 'cilia', or fine hairs, that form a central part of each taste bud. Generally, a molecule has to be dissolved in water to reach the cilia of the taste buds.

When we put food in the mouth, flavour molecules that are already dissolved in water (e.g. those in a sauce, etc.) are likely to reach the taste buds first and provide an initial taste sensation in the mouth. As we chew the food, so we release new flavour molecules into our saliva, also enzymes can start to react with proteins, etc. to produce new molecules through chemical reactions that actually take place in the saliva. Thus the taste sensation can change as we chew each mouthful.

Although we can only really detect the five distinct flavours, there remains a great deal of subtlety of taste in the mouth. We rarely taste foods that are purely bitter, sour, sweet or salty, so the different combinations of intensity of just these four tastes allow for a very wide range of tastes. If we consider only sweetness, different sugars taste more or less sweet; combinations of sugars can appear to be sweeter than either of the individual sugars. Most people find fructose to be sweeter than sucrose and glucose very much less sweet. So the actual sugar can affect the taste sensation.

When comparative tests are carried out different people have widely differing sensitivities to different sweeteners. So if you take two people both of whom like one teaspoon of sugar in their coffee, and ask them to use another sweetener instead, one may need only a tiny amount, while the other may need very much more. So we cannot be sure that any two people get the same taste sensation from any particular dish. The best any cook can hope for is that if they like the taste of a dish so will their guests.

The Sense of Smell

Our noses are much more discriminating than our tongues. We have 5 to 10 million olfactory cells sensing smells in our noses. We can detect the smell of some substances when only as few 250 molecules interact with just a few dozen cells.

The limitation of smell is that we can only detect air borne molecules. This limits us to smelling "small" molecules. Once there are more than a hundred or so atoms in a molecule it becomes too heavy to carry through the air in sufficient quantity for us to detect it by smell.

When we eat, most of the actual "flavour" is sensed in the nose. Each time we breathe some breath comes from the back of the mouth up into the nasal passages where it is sampled by the olfactory cells. The resulting sensation is the major part of what we call "flavour".

However, as with taste, the actual sensation varies with time. When we first put some food in our mouths only the most volatile molecules are carried in the air back into the nose where they are smelled. Generally it will be the smallest molecules that we smell first. Then as we chew the food more small molecules are released from the food and some of the larger molecules slowly evaporate into the nasal passages.

Some scientists, known as flavourists, spend a good deal of time and effort trying to understand the way in which different smells or odours are built up. They have managed to identify many hundreds of different chemical compounds that we can smell. They have also managed to analyse hundreds of different compounds in "simple" odours such as strawberries, with many more in more complex smells such as coffee.

However, these analyses, which have detected some molecules in concentrations as small as a few parts in a million or even less, are not usually adequate to allow the flavourists to reproduce the actual smells artificially. In practice, many of the compounds in a particular smell may be present in such low concentrations that even the most sophisticated scientific equipment is unable to identify them. However, our noses are able to smell them, and they often provide the key essential ingredients of a particular odour.

Not all people have equally sensitive noses. For example about a third of the population (about 40% of men and about 25% of women) cannot smell truffles. It is surprising for such an expensive ingredient that many who eat it simply have no idea what it tastes like (I have to admit that I am one of those who cannot smell truffles!). This variability in the sensitivity of people's sense of smell is probably one of the underlying causes of the wide diversity in food preferences. What smells quite repulsive to one person may seem attractive to another.

Where do flavours come from?

We can taste molecules of many sizes. There are some specific types of molecules we know we can taste. Acids will always taste sour, no matter what their size. Alkaloids are normally quite large molecules and always taste bitter. However, we can only smell comparatively small molecules that are swept up in the air into our noses. The overall flavour of a dish comes from the combination of both taste and smell sensations so is determined by a whole range of molecules. However, by far the greatest complexity is in the nose in the "smell" component of the flavour. So we can say that the majority of the flavour comes from the small molecules in the food.

Many foods have plenty of small flavour molecules present before they are cooked at all. Fruits all have characteristic flavours that are carried by small molecules. Of course, many plants want to advertise that they have fruit available. Animals then eat the fruit and distribute the seeds as they pass through the digestive systems unharmed. So the need for plenty of flavour to encourage consumption of the fruit is a good evolutionary trend.

Other foods have little flavour until they are cooked. Meats are made mostly of large protein molecules that have no smell and very little if any taste. However, as we will see in the next section, new small flavour molecules can be created in the cooking process.

Chemical reactions in cooking

There is a range of important chemical reactions, which help to develop flavour during cooking. The first group contains the enzymatic reactions. These are natural chemical reactions that affect the food. All foods contain enzymes – there are many different enzymes that are present in different foods – in nature these enzymes control biochemical reactions essential for the life of the organism. The reactions can continue once the organisms are being used as food. Examples include the ripening of fruit, the setting of cheese, and the breaking down of proteins in the ageing of meat.

To understand the enzymatic reactions in particular cases requires a good deal of biochemical knowledge and lies outside the scope of this simple introduction. A few special cases will be discussed in later chapters where they are particularly important.

The second group of reactions are those which affect sugars and carbohydrates when they are heated. Many disaccharide and oligo-saccharide sugars will undergo a process known as hydrolysis when heated with some water. The water reacts with the oxygen atom joining the sugar rings and breaks the complex sugars down into single ring sugars. The best known example is the conversion of sucrose to a mixture of fructose and glucose which occurs during the preparation of boiled sweets, etc.

If sugars are heated further, then additional reactions take place and the rings will open up to form new molecules – these reactions are generally called degradation reactions. Degraded sugars form acids and aldehydes.

If the temperature is increased to a sufficient temperature that sugars melt then more complex reactions occur that start to oxidise, or burn the sugar. These are called caramelisation reactions. The caramelisation begins with the conversion of sucrose to glucose and fructose as described above. The ring structures of these smaller sugars are then broken open through degradation reactions and these smaller molecules recombine to form chain like molecules. As the complex (and as yet little understood) reactions progress so the colour of the system changes from a clear liquid through yellow to dark brown.

During caramelisation a whole range of new small flavour molecules are formed. Many of these molecules have been identified as a range of organic acids that are formed along with the brown coloured polymers. As the reaction proceeds so the new molecules that form tend to be more like the alkaloids and have increasingly bitter tastes.

Probably the most important chemical reactions that occur during cooking are those that occur between proteins and sugars – these have become known as the Maillard reactions and deserve a section to themselves.

The Maillard reactions

Louis-Camille Maillard never worked on food. He was a Physician who concerned himself with the biochemistry of living cells. His work led him to investigate how the amino acids would react with sugars both of which could be found inside cells. Today, Maillard's name is inextricably linked with food science. You cannot pick up any textbook to do with food and fail to find his name in the index. The reason is simple. Long after Maillard died, it was realised that all the meaty flavours that develop during cooking are caused by reactions of amino acids with sugars. Maillard's pioneering work in the area led to the whole group of complex reactions being given his name.

The reactions are very complex and the details are by no means understood today, despite many chemists devoting their lives to study the reactions. The complexity comes from the fact that there are many different sugars and amino acids that can react together and from the fact that the actual reaction products from any one sugar amino acid pair depend on the temperature at which the reaction takes place, the acidity of the environment, the other chemicals that are nearby as well as random chance!

However, this very complexity offers the chef a range of interesting possibilities. Nearly all the molecules generated in the Maillard reactions (so far well over a thousand have been identified) are volatile enough to count as "flavour" molecules. So from the same starting ingredients the control of the temperature and environment can lead to a range of differing flavours.

In the Maillard reactions the amino acids can come from any proteins and the sugars from any carbohydrates. In the first stage of the reactions the proteins and carbohydrates are degraded into smaller sugars and amino acids. Next the sugar rings open and the resulting aldehydes and acids react with the amino acids to produce a wide range of chemicals. These new molecules then react amongst themselves to produce the main flavour compounds. The list of identified compounds includes several important classes of molecules. Pyrazines are the molecules that give fresh green notes to fruit and vegetables. Furanones and Furanthiols have fruity odours. Other compounds such as the di sulphides have pungent and even unpleasant smells.

One particularly important molecule that is generated has been associated with meaty odours – if it is missing the meaty smell is absent. When present even in very small quantities there is a strong meaty smell. This molecule is called bis-2-methyl-3-furyl-disulphide and is now widely used in the flavour industry to prepare artificial meat flavours.

Controlling the Maillard reactions is a tricky business – really it is a part of the art of a good chef to know how much heat to apply to a piece of meat to get the flavour he wants. There are however, a few simple guidelines that can be helpful. The Maillard reactions only take place at all quickly at high temperatures (above about 140°C) so you need to cook meats at high temperatures to develop "meaty" flavours. Since these high temperatures will only occur at the surface of the meat (inside there will be water which cannot be heated above 100°C without turning to steam) you will develop the flavour more quickly if you increase the surface area of the meat. You can increase the surface area by cutting the meat into small pieces, or thin slices before you cook it.

A second important point to bear in mind is that as the temperature rises above about 200°C, so new molecules start to appear at the end of the Maillard reactions. Some of these molecules that are formed at high temperatures can be carcinogenic and don't taste very pleasant. So it is wise to avoid over-heating your meats. The potential for dangerously over-heated meats is greatest on barbecues where some cooks may completely carbonise the outside of the food!

Experiments to try for yourself

Two experiments to show flavour is all about smell

Tasting Crisps

For this simple experiment you will need a friend or two to act as guinea-pigs, blindfolds for each friend, several packets of crisps with different flavours (one pack must be of plain or ready salted crisps) and a few other foods with strong smells (e.g. fruits and cheeses, etc.).

Begin by blindfolding the guinea-pigs. Tell them you will give them some crisps to taste and they have to identify the flavours. Now take a plain crisp and one of the flavoured crisps. Put the plain crisp in the guinea-pig's mouth and at the same time hold the flavoured one just under their nose – making sure it doesn't touch them so they don't realise what you are doing. Tell the guinea-pig to eat the crisp and describe its flavour. Repeat feeding your friend another plain crisp but holding other crisps and even the smelly foods under their noses.

You should find that the guinea-pigs reckon they are eating crisps with flavours of the ones you hold under their noses. If you hold a strawberry under their nose they may even tell you they have a strawberry flavoured crisp in their mouth!

This experiment demonstrates how most of our perception of flavour comes from the nose and not the mouth. The trick is not perfect as there are real differences in the way in which we perceive smells depending on whether we breathe the molecules in through our nose or whether they enter the nasal tracts from the back of our mouths.

A variety of purees

These experiments demonstrate the limitations of our tongues to tell different flavours apart. You will simply feed several different pureed foods to your friends while they hold their noses so no flavour molecules can be carried up into the nose and be smelled there. You will need a friend or two to act as guinea-pigs (they can test you later as well!), blindfolds so they cannot tell the different foods apart from their colours (alternatively you could dye all the purees the same colour using a food dye). If you have access to divers nose clips they will prevent any cheating when people hold their own noses.

Prepare a series of purees, three from fruit: e.g. apples, pears, peaches; three from vegetables: e.g. potatoes, peas, carrots and three or more from other distinctive foods: onions, tomatoes, baked beans and, if you dare, garlic.

Now get the guinea-pig to taste the purees in any order and identify which is which (you can tell them what the possibilities are if you wish or just leave them guessing).

The fruits will all taste a little sour from their acids so most people should be able to identify them as fruits – but are very unlikely to be able to tell which fruit is which. The vegetables have few distinguishing features (except carrots may be a little sweeter than the others) and will just taste rather bland. The onions can excite the trigeminal sense. This is the sense that is activated by "hot" foods such as curries and chillies – the tingly sensation at the back of the nose – some people will be able to identify it (however, if you don't say one of the foods is onions they are just as likely to say pepper as onion!). The tomatoes will excite the Umami taste buds and should be identified easily – however, if you get some monosodium glutamate from your local Chinese store or health food shop, it is likely to be confused with the tomato. Baked beans contain lots of sugar and lots of salt as well as tomato sauce – normally the salt and sugar counteract one another leaving just the Umami taste from the tomato puree. However, in this test some people will taste the saltiness and others the sweetness – so the baked beans can be mistaken for just about anything!

Heating and Eating – Physical Gastronomy

Why Do We Heat Our Food?

Cooking food increases the range of foods we can eat. Foods which would otherwise be indigestible can be made edible. For example we cannot digest raw potatoes, since the starch is in a form our stomach is not capable of processing; however, by heating to a high enough temperature the starch is altered, and it becomes edible. Some foods may contain toxins (e.g. pork), these toxins can often be destroyed by the application of heat. Thus cooking food can lead to a reduced risk of food poisoning.

Cooking can also change the texture of foods; for example the 'tenderising' of some meats can make otherwise unappetising food more appealing, and increase the available food supply.

Further, as we saw in the previous chapter, cooking often leads to chemical reactions that change the flavour of foods by breaking down large molecules (which we cannot taste) into smaller molecules that we can taste.

Heat and Temperature

Most people, if they even think about it at all, don't think there is much difference between heat and temperature. However, they are two entirely different concepts and understanding the difference will really help to improve your appreciation of the different methods of cooking.

Heat is the energy that flows from a hot body to a cold one. The temperature is a measure of which way heat will flow. When two objects at different temperatures are brought together, heat will always flow from the object at the higher temperature into the object at the lower temperature. As heat flows into the colder body so its temperature rises, while as the heat flows from the hotter object its temperature falls. Eventually the two objects reach the same temperature and no more heat flows.

If you put a cold dish in the oven heat flows from the air in the oven into the dish and its temperature rises until it reaches the temperature of the oven. The temperature in the oven is maintained as heat flows out of the air into the dish, by the heating elements in the oven. Similarly, if you put a hot pan in the fridge, then it cools down to the temperature inside the fridge as heat flows from the pan into the fridge (and is then extracted by the action of the refrigerator).

Suppose you put two different objects both at the same temperature, perhaps 20 °C (for example, a piece of metal and a bowl of water at room temperature) in an oven at a fixed temperature of say 50 °C. Heat from the oven will flow into both objects and they will both heat up. But the rates at which they heat up will not be the same. The amount of heat that must flow into each object to raise its temperature from 20 °C to 50 °C will differ for different objects.

Scientists define the "specific heat" of a substance as the amount of heat required to raise the temperature of 1kg of the substance by 1 °C. Metals tend to have much higher specific heats than water. So in the example above, the piece of metal would need to absorb more heat than the bowl of water (assuming they each had the same weight). In practice, the metal would take longer to heat up to the temperature of the oven.

Latent Heat

There is an exception to the idea that if we add heat to a substance its temperature will increase. This is when the material changes state, from a solid to a liquid, or a liquid to a gas. We all know that when we put a pan of water on the stove to heat up, its temperature rises until it reaches 100 °C and then it starts to boil. The temperature of the water remains at 100 °C until it all boils away, but we keep putting in more and more heat to make the water boil.

In fact it takes a lot of heat to change the state of water from a liquid to a gas. We call this heat "Latent Heat"; specifically for boiling water it is the "latent heat of vaporisation of water".

If we allow steam to cool down and condense (as for example when cooking in a steamer) it has to give up this latent heat to change from a gas to a liquid. This is one reason why steam burns are so severe. The amount of heat steam will put into your skin is higher than the amount of heat the same weight of hot water will put in your skin. The steam has to dump its latent heat to turn from a gas to a liquid before it can cool any more.

Exactly the same concepts apply to changing state from a liquid to a solid. When we freeze ice, we have to remove a very large amount of latent heat. This is why it takes a long time to freeze water. Conversely, ice cubes last a reasonable time in our drinks as they have to absorb large amounts of latent heat before they can melt.

Which freezes more quickly; hot or cold water?

A few years ago, a schoolboy in Africa, Erasto B Mpemba, made an important observation. He noted that if he put two identical containers each with an identical amount of water (but at different temperatures) in a freezer side by side, the container of initially hot water always froze before the container of colder water. He persistently asked his Physics teachers why this should happen and was repeatedly told he was wrong – it can't happen.

The argument of his physics teachers was that the hot water would take some time to cool down to the temperature the cold water started at by which the initially colder water would have cooled to a lower temperature. Accordingly the hot water will always be hotter and cannot freeze first. Erasto, however, was a very persistent lad and despite being teased about his insistence that hot water freezes faster than cold water he never stopped believing the results of his own carefully conducted experiments.

Eventually, a British Physicist, Prof Osborne, was invited to speak at Erasto's school and talked for 30 minutes or so about physics and national development. Students asked him about gaining University entrance and other topics. But Erasto asked "why does hot water freeze faster than cold water?" Fortunately, Prof Osborne is a sensible Physicist and asked Erasto what he meant – Erasto explained his experiments to Prof Osborne who was very surprised as the results contradicted his intuitive grasp of thermal physics.

Prof Osborne promised he would repeat Erasto's experiments when he returned home, and confirmed Erasto's findings. If you find it hard to believe, then try it for yourself.

Although the effect is now well known, it is not fully understood. A variety of explanations have been proposed, including: the idea that some evaporation of the hot water leads to an additional cooling effect; the concept that the convection currents caused by the high temperature difference of the hot water from its surroundings improve the rate of heat transfer; the suggestion that boiling the hot water precipitates some small impurities that act as nucleating sites for ice crystals; and the idea that a hot container when put in a freezer can melt some ice on the surface it is in contact with leading to a better thermal contact with its surroundings. As far as I am aware, none of these explanations on its own is sufficient to explain Erasto's simple observation.

Heat flow

The rate at which heat is transferred from one object to another depends on many things – how well the two objects are in contact, how fast heat can flow through the objects, the specific heat of each object and importantly the difference in the temperature of the two objects.

When a hot body is in contact with a colder body heat will flow from the hotter to the colder body. The rate at which heat flows depends on the difference in the temperatures of the two bodies. If we write the rate at which heat flows as H, and the temperature of the cooler body as T_L and that of the hotter body as T_H, then H is proportional to $T_H - T_L$. A scientist would write this as:

$$H \propto (T_H - T_L)$$

Of course as heat flows so the temperature of the hot body decreases and the temperature of the cold body rises, so the difference in temperature $(T_H - T_L)$ is reducing all the time. So the rate at which heat flows decreases as the two bodies approach equilibrium.

Of course, other factors also affect the rate at which temperatures of the hot and the cold bodies approach one another. For example, if the cold body has a high specific heat so that it requires a lot of heat to increase its temperature, then its temperature will rise only slowly as heat flows into it. The rate at which heat will flow through a body is also important. Here the concept of thermal conductivity is important. Some materials are good thermal insulators – they resist the flow of heat. We use thermal insulators to make oven gloves that allow us to handle hot pots and pans. Other materials, such as many metals conduct heat very well so that if one end of a metal spoon is put in a pan of boiling water the other end quickly heats up and becomes too hot to handle. The high thermal conductivity of most metals is why we use wooden or plastic spoons to stir saucepans. The exception is stainless steel – stainless steel has a low thermal conductivity so kitchen implements made from stainless steels are very useful (if expensive).

Methods of heat transfer

In all the above, we have not considered the actual process of how the heat moves from the hot to the cold body. There are several distinct ways in which heat can be transferred to an object each with its own characteristics and all used in cooking.

Conduction

The transfer of heat inside any solid occurs by conduction. Consider a brick being heated from the bottom. Initially the brick is at the temperature of its surroundings. Once the brick is heated from the bottom the temperature of the bottom surface will increase and heat will flow into the brick. This heat will be conducted through the brick at a rate depending on the thermal conductivity of the brick. As heat is transferred into the brick, so it will heat the brick leading to a temperature gradient inside the brick. The temperature inside the brick will be highest at the bottom and will fall towards the top. As time passes and more heat is put in the brick so the distribution of temperature inside the brick will change.

As heat flows into the brick, so it heats the brick up raising the temperature locally. The actual temperature rise depends not only on the thermal conductivity of the brick, but also on its specific heat – if it has a low specific heat then the temperature increase will be larger than if it has a high specific heat. It becomes convenient to define a new material property, the "thermal diffusivity" that combines the specific heat, the thermal conductivity and the density of the material. The local temperature rise as heat is conducted through the brick then depends on this thermal diffusivity.

In cookery we will usually be interested to know about the time it will take for the temperature at the coolest part of the food to reach a certain value. The solutions of the heat flow equations can be quite complex, but there are certain features that always come up and which can be visualised fairly simply.

Thermal Conduction – a Mathematical Treatment

Energy in the form of heat can be passed along a material-atoms in one plane vibrate more rapidly as they become hotter and then, through collisions and other interactions pass on some of this increased energy to atoms in the next plane and so on.

If we observe how heat is transferred through a thin slab of homogeneous material we can begin to find experimental relationships with which to build up a "theory" of thermal conduction.

We find the rate at which heat is transferred across our slab, H ($= Q/\Delta t$), (where Q is the amount of heat transferred in a time, Δt), depends linearly on the temperature difference, ΔT, across the slab. That is: $H \propto \Delta T$.

We find that the rate of heat flow depends inversely on the thickness of the slab (Δx). That is: $H \propto 1/\Delta x$.

We find the rate of heat flow depends linearly on the area, A, of the slab (i.e. $H \propto A$). So we can write:

$$H = \frac{Q}{\Delta t} \propto A \frac{\Delta T}{\Delta x}$$

we can then define a constant, k, the thermal conductivity such that:

$$H = kA \frac{\Delta T}{\Delta x}$$

If we consider a simple case of a conducting rod connecting two thermal reservoirs maintained at constant temperatures, T_H and T_L, if the rod is insulated so that no heat is lost from the sides then heat will flow along it from the hot reservoir to the cold reservoir. And the temperature in the rod will decrease linearly along its length from T_H to T_L.

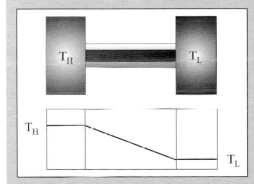

This is a steady situation, the temperature in the rod remains the same at all times.

To look at problems where the geometry is more complex, or when there is not a steady state, it is often simpler to use a differential form of the equation for thermal conductivity:

$$H = \frac{dQ}{dt} = -kA \frac{dT}{dx}$$

the minus sign appears because the heat flows in the direction of decreasing temperature.

As heat is conducted into an object it will take some time for it to reach a distance, x inside (this will depend on the thermal conductivity and the initial temperature difference between the outside and the inside). The heat conducted into the object heats the material so raising its temperature (this temperature rise depends on the specific heat and the density of the material). The result of combining the time it takes heat to flow through the object and the time it takes that heat to raise the temperature is that the total time taken for the temperature to rise by some specified amount at a distance, x from the outside of a body, is proportional to x^2. Alternatively we can say the cooking time depends on the square of the size of the dish we are cooking.

Convection

As well as heat flowing through a body, from the outside to the inside, heat has to be transferred from the surroundings to the object. Convective heating is the transfer of heat from a fluid (a gas or a liquid) to its surroundings. The fluid is kept in motion and is able to take heat from the "source" (the hot body, or a source of heat such as a fire) to the "sink" (the cold body). Convection is the commonest form of heating used domestically. We heat our houses by convective heating. Water is heated by a central boiler and pumped to "radiators" in the rooms to be heated. The radiators containing hot water heat the air around them. The hot air rises and sets up a flow through the room circulating the hot air to warm the surroundings and eventually return, having cooled down, to the radiator where it is heated again.

In cookery we use convective heating in many situations. For example, boiling or simmering in a pan of water – the water is heated at the bottom of the pan on the stove and the hot (or boiling) water circulates around the pan heating the contents. When we cook in a deep fryer, the oil is heated and circulates around the food transferring heat in the same way. We use gasses as well as liquids as the fluid to transfer the heat. In an oven the air is heated by the elements (or by burning gas, etc.) and circulates around the oven heating the food inside. Some ovens have a fan to force the air to circulate more frequently, others rely on the natural circulation caused by the heated air rising and the cooler air falling as in the domestic heating systems.

In general in a convective heating system used in cooking, we maintain the heat transfer fluid at a fixed temperature by applying heat to the fluid through an electric heater, or by burning gas. When we use water as the heat transfer fluid we can keep it at a constant 100 °C simply by keeping the water boiling. When we use other fluids (oils in a deep fryer, or air in an oven) we have to use some other form of control – generally we use a "thermostat". A thermostat is a device that automatically turns on the heat source when the temperature of the fluid drops a few degrees below the required temperature and turns the heat source off again once the fluid has heated back to the required temperature.

In most domestic equipment these thermostats are very inaccurate – so it is well worth investing in a thermometer to check the actual temperature of your

Thermal diffusivity and the heat transfer equation

We can use the differential form of the equation for thermal conductivity to see how heat actually diffuses into a substance. We simply state that the sum of the overall heat entering a small volume element in a small time is equal to the specific heat of the material multiplied by the temperature change in that small time interval.

The net heat entering a small volume element of area A and thickness (x, in time (t is given by:

$$-\frac{d}{dx}\left(\frac{dQ}{dt}\right)\Delta x \, \Delta t$$

this must be equal to the temperature change in the same time, i.e.:

$$-\frac{d}{dx}\left(\frac{dQ}{dt}\right)\Delta x \, \Delta t = CA\,\Delta x \,\frac{dT}{dt}\,\Delta t$$

$$-\frac{d}{dx}\left(\frac{dQ}{dt}\right) = C\frac{dT}{dt}.$$

But the equation for thermal conduction gives:

$$H - \frac{dQ}{dt} = kA\frac{dT}{dx}$$

if we substitute this in the continuity equation above we find:

$$-\frac{d}{dx}\left(\frac{dQ}{dt}\right) = kA\frac{d}{dx}\left(\frac{dT}{dx}\right) = kA\frac{d^2T}{dx^2} - CA\frac{dT}{dt}$$

i.e.

$$\frac{dT}{dt} - \frac{k}{C}\frac{d^2T}{dx^2} = \kappa\frac{d^2T}{dx^2}$$

(κ is known as the "Thermal Diffusivity").

To solve this equation we need to put in the "boundary conditions" for a particular geometry of the object being heated and the circumstances under which heat is being transferred. There are many particular solutions, but generally, we can only solve the equation approximately to find the distribution of temperature inside our object as a function of the time it is heated for.

Now although the thermal diffusion equation above has many solutions, in all of them a term of the form x^2/t appears. For example, one solution is:

$$T(x, t) = (4\kappa t)^{-1/2}\exp\left(-x^2/4\kappa t\right)$$

Thus the cooking time is always proportional to the square of the size of the food, rather than its weight.

oven, etc. and to see just how much it varies as the thermostat switches the heat on and off. A good forced convection oven (one with a built in fan) will keep the circulating air at a temperature within 5 °C of the set temperature at all times. A good oven that relies on the natural convection currents will have a distinctly higher temperature at the top than the bottom. However, at any one place inside the oven the temperature should not vary by more than a few degrees as the heater switches on and off.

Radiation

All bodies, when hot, radiate heat. The amount of heat they radiate depends on the fourth power of their absolute temperature (to find the absolute temperature in Kelvins, K, take the temperature in °C and add 273). For example, the sun is at a very high temperature, 5800 K, and radiates a great deal of heat to the earth. We can feel how this radiation warms us as we absorb it on our skin on a sunny day.

Some of the heat radiated from the sun is absorbed by the Earth, which slowly heats up. As the temperature of the earth rises, so it radiates more heat back into space. Eventually the rate of radiation of heat into space equals the rate of arrival of heat from the sun and a steady state is achieved. This leads to a mean surface temperature of the earth of around 300 K.

However, if we increase the amount of heat absorbed at the Earth's surface, from the sun's radiation, then the surface temperature will rise until a new equilibrium is reached. In practice, the less ozone there is in the Earth's upper atmosphere the greater the amount of radiation from the sun that can reach the surface (UV radiation normally does not reach the surface).

In cooking we use radiation in two ways, when we grill our food the heat radiates from the grill and is absorbed at the surface of the food; microwaves are also a form of radiation and are absorbed by water in the food.

Applications to cookery

In cookery, we do not simply bring two bodies into contact, rather we put some food (say a potato) into a pan of boiling water, or a hot oven. We apply heat to the water or the oven to keep it at a constant temperature. The food then heats up and physical and chemical changes take place. We say the food is cooked when it has been at some high temperature for some specified time. If we have a little understanding of the way in which heat flows into a body, we can calculate these "cooking times" for any situation.

Methods of Heating in Cookery

Cookery books are usually rather imprecise when talking about heat and temperature. For example terms such as cook on a low heat are often used. It can be difficult to work out exactly what is meant. Generally it depends on the context; so if one is baking a low heat means an oven set at a relatively low temperature (say 140 to 160°C). If the dish is being cooked in a pan on the top of the stove, then a low heat should be interpreted as meaning reduce the power of your burner to the lowest possible setting.

In the following sections I will describe the physics behind the different ways in which we heat our food in the kitchen, in the hope that a little understanding may assist in interpreting the sometimes vague instructions in recipes.

Convection

Convection is the main process of heat transfer which occurs in many cooking methods. Boiling, baking, and deep frying are all examples of cooking using convective heat transfer. But we all know from experience that cooking times vary between these different methods. Why?

We can begin to understand the variations in cooking times by looking at one (approximate) solution of the equation governing heat transfer for a sphere of radius, r:

$$r \propto \kappa s \log(T - T_c) \sqrt{t},$$

where κ is the thermal diffusivity of the food, s is the specific heat of the surrounding medium, T the temperature of the cooking medium, T_c the required final temperature at the centre of the food, and t the cooking time. We will find it easier if we rewrite the equation it to give cooking times this gives:

$$t \propto \frac{r^2}{\kappa^2 s^2 \left[\log(T - T_c)\right]^2}$$

The cooking time is proportional to the square of the smallest radius of the foodstuff, r, and inversely proportional to the square of the difference between the temperature of the surrounding medium and the required cooking temperature. Also note that higher specific heats of the surrounding medium will rapidly decrease the cooking time.

Now the specific heat of air is very much less than that of water. We can happily put our hand into a hot oven for a few seconds without being harmed, because the amount of heat transferred to our hand depends on the specific heat of the air and is not enough to heat our skin too much. Conversely, if were foolish enough to put a hand into hot water, it will be scalded; the specific heat of water is comparatively high and our skin will rapidly be heated to the temperature of the water, and damaged.

Thus we can see why cooking something like potatoes takes longer in an oven than by boiling, even though the oven may be at a higher temperature. However, when we use deep fat frying, the cooking time is reduced as compared to boiling. This happens since there is a much higher temperature of the fluid (typically 180 °C compared with 100 °C).

Example – boiling an egg

Each egg is unique. Eggs vary in shape and size as well as in the relative size of the yolk and white, etc. Cooks who find procedures that work stick to those procedures come what may and swear it is the only way to guarantee perfect boiled eggs every time.

Not every solution will work for everybody. We are all individuals with different preferences for the texture of our eggs. Recipes may only work with the actual equipment on which they were designed. Several popular food writers suggest putting the egg in boiling water for a short (specified) time and then turning off the heat and allowing the egg to continue to cook in the hot water for a further (specified) time. However, the degree to which the egg will cook depends not only on the shape and size of the egg itself, but also on the actual temperature of the water in the pan as it cools. In turn, this temperature depends on the material the pan is made from, the shape and size of the pan, the amount of water in the pan, where you leave the pan to stand, and even the temperature of your kitchen.

A little understanding of what actually happens when you boil an egg may help you to adapt any particular technique to suit your own equipment and eggs of more or less any size.

Inside the shell, eggs have two main parts, the white and the yolk. Both the white and yolk contain proteins that will, when suitably heated, undergo chemical reactions to make them solidify. Controlling these reactions allows the cook to control the firmness of the white and yolk in a boiled egg.

As eggs cook, so the proteins first denature and then coagulate. Little happens until the temperature exceeds some critical value and then the reactions begin and proceed faster as the temperature rises. In the egg white the proteins start to coagulate once the temperature is above 63 °C, while in the yolk the proteins start to coagulate once the temperature is above about 70 °C. For a soft boiled egg the white needs to have been heated for long enough to coagulate at a temperature above 63°C, but the yolk should not be heated above 70 °C.

The 'runniness' of the yolk depends on the extent to which the yolk proteins have coagulated – i.e. on the temperature the yolk has reached and the time it has been kept hot. In practice, the rate of the reactions increases so fast with increasing temperature, that it is sufficient to say the egg is cooked perfectly when the centre of the yolk has just reached some particular temperature. The actual temperature required depends on the individual's taste. For those who prefer runny eggs the temperature should be well below 70 °C, for a firmer yolk the temperature at the outside of the yolk can be allowed to approach 70 °C during the cooking process – heat will continue to diffuse from the white into the yolk as the egg stands on the table, so firming up the yolk. For a very firm yolk, or a hard-boiled egg, the centre of the yolk should be allowed to reach 70 °C before removing from the boiling water.

Cooking the perfect egg is thus simply a matter of controlling the temperature inside the yolk (and white). The problem is to control these temperatures: a "simple" heat transfer problem that any undergraduate physicist should be able to solve.

Temperature	Effects on Egg-white	Effects on Yolk
Below 55 °C	Risk of Salmonella	
Up to 63 °C	Soft and gelatinous similar texture to partially set jellies or non-drip paints	'runny' liquid with similar "thickness" to washing up liquid
65–70 °C	Set as a soft gel – similar texture to well set jelly	Still runny, but liquid starting to thicken – viscosity increasing to that of treacle
73 °C	Hardening of white, texture of soft fruit, strawberries, etc.	Soft gel like texture rather like a thick shampoo
77 °C	White continues to harden	Hard boiled still soft, but solid texture of set yoghurt
80 °C		Onset of green discoloration around edge of yolk
90 °C	Overcooked tough white, texture of damp sponge.	Yolk completely dry, crumbly solid

After a few simplifications and approximations have been made it can be shown that the time taken for the centre of the yolk to reach a particular temperature, T_{yolk}, is given by:

$$t = 0.0015\, d^2 \log_e \left[\frac{2\,(T_{\text{water}} - T_0)}{(T_{\text{water}} - T_{\text{yolk}})} \right],$$

where d is the diameter of the egg in millimetres, T_0 is the temperature of the egg before it was put in the water (in °C), and T_{water} is the temperature of the boiling water (in °C) (as was first published in 1996 in the New Scientist by Dr Williams of Exeter University).

The most important feature of this equation is that the cooking time increases with the square of the diameter of the egg. A fairly small egg (with a diameter of less than 40mm) will take only about 60% the time to cook of an extra large egg (with a diameter of about 50 mm). The other important feature is that the cooking time depends on the temperature of the egg when it is put in the boiling water. An egg taken from the fridge at 4°C needs about 15% longer cooking time than an egg which starts at room temperature (20°C).

If you have suitable equipment you can cook eggs at lower temperatures – this will lead to a more uniform temperature through the egg and will produce a firmer white as it will have been hot for a longer time. For example a 45 mm diameter egg cooked in water that is kept at a constant 70°C for 8 minutes will have a very firm white and a runny yolk that has just reached 63°C in the middle.

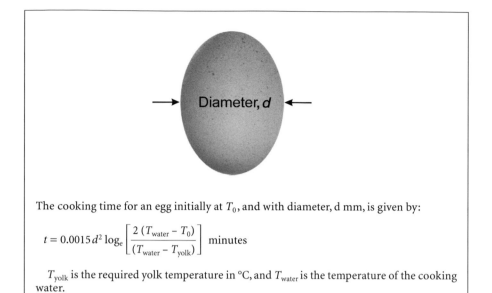

The cooking time for an egg initially at T_0, and with diameter, d mm, is given by:

$$t = 0.0015\, d^2 \log_e \left[\frac{2\,(T_{water} - T_0)}{(T_{water} - T_{yolk})} \right] \text{ minutes}$$

T_{yolk} is the required yolk temperature in °C, and T_{water} is the temperature of the cooking water.

Radiant Heating

When we use a grill to cook food most of the heat comes not from the surrounding air, but it is absorbed at the surface of the food as radiant heat. The interior of the food is then heated by conduction of the heat from the hot surface. This conduction inwards from the surface follows the same pattern as for convective heating, so the cooking time still depends on the square of the smallest dimension of the dish. However, the temperature of the surface is no longer determined by the temperature of the surroundings, but is determined, mainly, by the amount of heat energy falling on the surface, which in turn depends on the power of the heat source (the grill) and the distance between the grill and the food.

Broiling

Broiling, is a cooking method where food is placed in direct contact with a hot surface, and the main heat transfer is from this surface into the food. It is similar to shallow frying. Again we may note that heat will be transferred into the interior of the food by conduction, and that the cooking time will depend on the square of the size of the dish. However, as with grilling, we are heating only from one surface, so the relevant size is the thickness of the food on the hot surface.

Cooking Times for Roasts, Eggs, etc.

Just for fun I have included some of the details of the physics here. Please feel free to skip over all the equations etc. if you like. However, I do think it is worthwhile to realise that there is some really solid science involved when we look at methods of heating.

For the mathematically minded, I have written out the solution of the heat transfer equation for the case of a spherical object immersed in a heat bath kept at a constant temperature. For the sake of simplicity, we approximate all food to be spherical! – don't worry, it really doesn't affect the conclusions which are all that matters.

It can be shown that in a sphere of radius, a, the temperature inside the sphere, $\Delta T(t, r)$ is given by:

$$\Delta T(t, r) \approx \frac{2ka\Delta T_0}{3r} \sum_{n=1}^{\infty} e^{-\kappa \alpha_n^2 t} \frac{k^2 a^4 \alpha_n^4 + 3(2k+3)a^2\alpha_n^2 + 9}{k^2 a^4 \alpha_n^4 + 9(k+1)a^2\alpha_n^2} \sin r\alpha_n \sin a\alpha_n$$

where $T(t, r)$ is the temperature inside the sphere as a function of time, t, and radius, r; expressed as the difference between the temperature of the heat bath, and the local temperature. ΔT_0, is the initial temperature difference between the sphere and the fluid in the heat bath; is the thermal diffusivity of the sphere, and k is the ratio of the specific heats of the sphere and the fluid in the heat bath,
and the $\pm \alpha_n$, $= 1, 2, 3 \ldots$ are the roots of:

$$\tan \alpha = \frac{3a\alpha}{3 + ka^2\alpha^2}$$

These equations provide the full distribution of temperature inside the spherical roast, or egg that we are cooking. However, we do not need the full solution to understand how long we need to cook for. We simply say that something is cooked when the centre has reached some particular temperature, T_c, say. The distance, x, from the outside which has reached T_c, is then given by:

$$\chi \propto \kappa s \sqrt{t} \cdot f(T - T_c)$$

where κ is the thermal diffusivity of the food, s, is the specific heat of the surrounding fluid, which is at a temperature, T, and t is the time. The function $f(T - T_c)$ depends on the difference between the temperature in the heat bath and the desired final temperature and takes a form that depends on the actual approximation we use to solve the above equations – usually it appears either as $(T - T_c)$ or as $\log_e(T - T_c)$.

Microwaves

Microwave ovens operate on a similar principle to radiant heating, however the radiation is of a longer wavelength than that used in conventional grills (which employ infra-red radiation), and can penetrate further into the food. The radiation has a frequency which is exactly tuned to a frequency at which water molecules can vibrate. Thus when a microwave is absorbed by a water molecule it transfers all its energy to the molecule thus 'heating' it. The 'hot' molecule can then transfer some of its energy to its neighbours, so that heat is transferred by conduction again.

Microwaves are able to penetrate inside water to a depth of about 1cm so that it is not only the surface that is heated, but a region of up to about 1cm thickness.

Cooking times then depend on the power of the microwave oven, the amount of food being cooked in the oven, and the size of the food (if larger than ca. 2cm).

Microwave ovens do not provide a uniform density of microwave energy, in fact there are many 'hot' and 'cold' spots inside a typical oven. This inhomogeneity of microwave ovens can be seen using some experiments with potatoes described later in this chapter. Briefly, if you put a potato in a microwave oven and cook on full power for about 40 seconds and then cut the potato open you will find finger like regions extending from the outside where the potato has been heated. This is in contrast to the uniform heating seen in 'conventional' cooking methods.

It is this inhomogeneity in heating inherent in microwave ovens which forces the use of 'standing times' to allow the temperature to equalise in foods after cooking in a microwave oven.

Some Experiments to try for yourself

Experiments with potatoes

Potatoes are very useful in trying to observe how heat is transferred within a cooking foodstuff, or dish. When we heat a potato, above ca. 60 °C, there is a visible change in the appearance; the potato changes from having an opaque white texture and becomes translucent. We will see later why this happens, but for now we shall use the phenomenon to follow the flow of heat into a cooking potato.

If we put some potatoes into boiling water for various lengths of time, and then remove them and cut them open, we will see the growth of a ring of translucent material from the outside inwards. the width of this region is the distance, x, in the equation above which has reached a temperature of 60 °C (or higher). If we measure x for a range of cooking times, and make a plot of x against , we will be able experimentally to justify the equation:

$$x \propto \sqrt{t}$$

N.b. It is much easier to see the ring of translucent material if you do *not* peel the potatoes.

An experiment to see how uniformly your microwave cooks

Cook a potato on the highest power for about 2 minutes and then cut it open to see where it has been heated above 60 °C. You will see some regions where the texture has changed to become translucent just as in the experiment above; the potato reached a temperature of above 60 °C in these regions. How are these regions arranged? Are there any regions which do not start at the surface? If you cook another potato do you see the same pattern of heating?

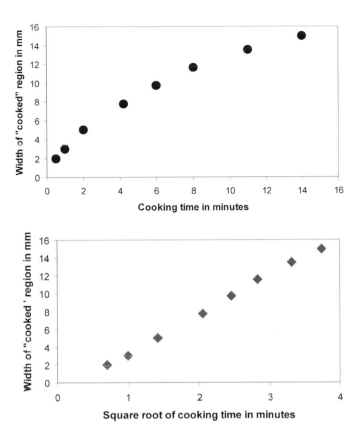

"Cooking a Light Bulb"

This experiment must only be performed under adult supervision and with the permission of the owner of the microwave oven – who should read the whole description before agreeing!

Microwaves are just a part of the electro-magnetic spectrum; they have a wave-length that falls between the infra-red and radio wave parts of the spectrum. We use microwaves not only to heat food, but also in telecommunications. The signals from mobile phones are carried by microwaves.

The mobile phones work by picking up the microwave signal in an aerial so that a small electric current is generated in the aerial. The electronics inside the phone then amplify this electric current and convert it into sound through the ear-piece.

Now once you realise that any piece of metal will act as an aerial (but it will act best if it has a size about the wavelength of the radiation) you can see that the filament in a light bulb will act as an aerial for microwaves.

You can demonstrate this very spectacularly if you take a domestic light bulb (say a 60 Watt bulb) and put it inside a microwave oven on the turntable. Now if you switch the microwave oven on **FOR A FEW SECONDS ONLY** the filament of the light bulb will pick up the microwaves and a current will be generated which will heat the filament (just as the current does when you switch on a light in the regular way) and the lamp will light up.

However, the microwave oven is capable of producing as much as 650 Watts of radiation, which is far more than the light bulb filament can take – so the filament is liable to melt! The heat generated can be quite enough to melt the glass envelope of the light bulb. If that happens the bulb will explode inside the microwave oven spreading broken glass around and making it essential the oven is carefully cleaned before any food is cooked!

It generally takes about 20 seconds on full power to explode a light bulb – so set the timer for a shorter time.

You can also use this demonstration to see how uniform the microwaves are inside the oven and to see how a microwave oven works when you use a lower power.

If you watch the light bulb rotating around the microwave on the turntable, you will see the brightness of the filament vary – it will be brighter where there is a higher concentration of microwaves in the oven and less bright where there are fewer.

If you use a lower power setting you will see that the microwaves are of the same power as on the high setting, but are switched on and off for different times. Typically at a 50% power setting the microwaves will be switched on for about 10 seconds and then off for 10 seconds, and so on.

Cooking Utensils Methods and Gadgets

Introduction

Today the cook is faced with a seemingly infinite array of kitchen equipment. If you can think of any task in the kitchen then someone will have invented a gadget to perform the task – whether it makes the job any easier is quite another matter! The choice of cooking equipment (ovens, hobs, etc.) grows every day with different fuels and heating methods regularly being introduced on the market. So how can you make the best choice?

A very important part of the scientific method is to ensure that whatever experiment you perform, you write about it in such a way that any colleague, anywhere in the world can reproduce the same experiment (and obtain the same result). When others fail to reproduce an experiment the original work is often discredited. A recent example is the work on "cold fusion" that received widespread publicity in the press before the original experiments could be repeated in different laboratories.

So if I were to write recipes in the same way as I write my scientific papers, I would need to be much more precise than is usual, and cooks using the recipe would have to use exactly the same equipment I prescribe in the actual recipe. The panel provides an example of how a scientific recipe might be written up. Clearly, it would be of little use to most domestic cooks!

All recipes are made up and tested with a particular set of equipment, in most cases they will only be truly reproducible when using identical equipment. So some idea of the ways in which different equipment affects the actual preparation and cooking processes may help you adapt any recipe to your own equipment.

A "Scientific" recipe for boiling an egg

Materials

Use eggs that are three days old. The egg should have its largest diameter between 55 and 58 mm and smallest diameter between 48 and 52mm. It should weigh between 70 and 75 grams. Use double distilled water as a heat transfer medium. The heating should be performed using a standard laboratory hotplate with a magnetic stirrer. A Eurotherm should be used to control the heating rate and heat input. The vessel used to contain the water should be a standard Pyrex 600ml beaker.

Method

Measure and weigh the egg. Only eggs fitting within the parameters described above are acceptable for further experimentation. Fill the beaker with 450 ml of double distilled water at 20°C and insert the magnetic stirrer. The control thermocouple from the Eurotherm should be fixed in the beaker 1 cm from the bottom and as near to the centre as possible. The egg (already at 20°C) should be placed in the water and the magnetic stirrer switched on. The hotplate should then be heated to provide a constant 30°C/min heating rate until the temperature reaches 100°C. The Eurotherm should then be used to control the heat input to the heater so that a constant temperature of 100°C is maintained with a loss of no more than 10ml/min of water from evaporation. The egg should be removed from the water 3 minutes after the temperature reached 100°C. The egg should then be placed directly into a standard egg cup and served on the table within 30 seconds of removal from the water. The egg should be eaten by the diner within three minutes of being served.

There are so many gadgets and kitchen tools that it seems pointless to do more than just to outline a few that I find particularly useful. These are tools I use regularly in the kitchen and which have been widely used in the preparation of the recipes described in later chapters. When it comes to cooking pots and pans, there are some important differences that relate to materials and the shapes of the pans that are worth some discussion. Finally, I will briefly discuss the relative merits of the different types of ovens and hobs, as I see them.

However, in all this discussion you should bear in mind that choice of cooking equipment is largely a matter of personal choice – what suits one cook another may find almost impossible. Probably the best example in my experience concerns a humble potato peeler. I have always used a traditional peeler with a metal blade bound with some string to a wooden handle – the sort that our grandparents doubtless used. Some years ago, my parents bought a modern device that they swore by – indeed I have seen a similar tool in use in many kitchens. This peeler consists of a blade that pivots in a U-shaped handle and is supposed to peel in any direction. I have seen several cooks wield this tool with aplomb, but in my hands it never seems to work at all!

Tools and Gadgets

Hand Tools

Every kitchen needs a liberal supply of cutlery. When stirring hot foods you need tools made from insulating materials. Wooden or stainless steel spoons are particularly useful.

All cooks will need a range of mixing bowls – I have plastic, pyrex glass and metal bowls and have my own favourites for different jobs. There are no particular reasons for most choices apart from size. However, when whisking egg-whites by hand, a copper bowl can have some advantage – besides the possible effect of copper ions helping to coagulate proteins, the excellent thermal conductivity of copper and the fact that you hold the bowl under your arm as you whisk helps to warm the egg-whites, which in turn makes the denaturing stage require a little less effort!

Amongst the other essential items are chopping boards. These need to be made from a material (e.g. wood or plastic) that is much less hard than your knives, otherwise your knives will quickly become blunted.

There is much argument about the use of wooden utensils. In Britain, the use of wooden utensils is not generally permitted in professional kitchens as it is argued that they can harbour bacteria and may be a health hazard. However, several reports suggest that the enzymes present in wooden chopping boards can actually kill bacteria, while plastic boards may have a significant bacterial population. Personally, I have used wooden spoons and both wooden and plastic boards for many years and found no problems with hygiene.

Knives

To me the most important tools in the kitchen are my knives. With a sharp knife it is possible to prepare meat and vegetables quickly and safely. I prefer heavy stainless steel knives – while stainless steel knives are a little more difficult to sharpen, and hold their sharp edges perhaps for less time than carbon steel knives they have one major advantage – you can put them in a dishwasher! I use three sizes – small knives (with a blade about 10 cm long) for preparing vegetables and fine work, a medium cooks knife (with a blade about 16 cm long) for general chopping, and cutting meats, etc. and a large cooks knife (with a blade about 23 cm long) for any heavier duty work.

The heavier a knife is the easier it is to slice uniformly, so I tend to use the largest knife that is suitable for any job.

However, the most important consideration when using any knife is that it should be really sharp so that little effort is required to cut through your food. If you use a lot of force then the risk of the knife slipping and causing injury is much greater. Accordingly, a most important gadget to keep in the kitchen is a knife sharpener of some type.

There are a range of sharpeners available and most work well. I am very fortunate as my partner is a keen wood-turner so we have a water grinding wheel in our basement which I use to keep all my knives sharp. However, most knives will benefit from a little touching up of their cutting edges before each use – this is where the steel comes in to its own. Many people have a steel as a part of a carving set and never use it. It can take a little practice to master the use of a steel, but the results are well worthwhile. The simplest way to use a steel, is not to try any fancy movements but to hold the steel vertically against a chopping board with one hand and then to run the knife to be honed up the steel keeping an angle of about 30° between the knife blade and the steel. As you run the knife up the steel, so you draw it towards yourself thus honing its entire length in each pass. A couple of passes are all that should be needed to keep your knives very sharp.

Juice Extractors and Zesters

When using fruits, especially citrus fruits, you will often need to extract the juice and scrape the zest from the peel. I have two types of juice extractor – the traditional glass dish with raised ribs on which you twist the halved fruit and a much more efficient machine that is really a hand operated press that squeezes all the juice from any fruit. These days I only use this hand press and the glass dish remains lost at the back of a cupboard somewhere.

Thermometers

As cookery relies so much on heating, it is a mystery to me why so few cooks regularly use a thermometer. I have several. The most useful are a glass "jam thermometer" and an electronic thermometer with a probe I can use more or less anywhere.

I find using a thermometer essential when preparing custards, and similar dishes that require egg proteins to just begin to coagulate, but where over-heating will lead to undesired textures. Thermometers are also helpful when cooking meats, cakes, soufflés and chocolate. In all these dishes knowing the actual temperature inside the food will greatly assist in producing consistently excellent results every time.

Power Tools

Mixers

I find my power whisks very useful mainly when making cakes, but also occasionally with sauces, etc. I have two kinds: a small hand held whisk which is adequate for most general purposes and a large table top unit which is needed only when I am making large cakes, etc. This larger mixer comes with a range of attachments including a really useful blender – much used in sauce making and a mincer which is hardly used at all as the food processor is generally better.

Food Processors

My food processor is used almost every day. It cuts, shreds, grates or chops anything. The best advice I can give is to buy the most powerful processor your budget will allow. Food processors do a lot of work and the more powerful the motor the easier it will be for the blades to do the work you are asking of them and the longer the machine will last. So far I have managed to wear out two food processors in about 20 years!

Dishwashers

Before I had a dishwasher I always thought they were an unnecessary luxury. However, within weeks of owning a dishwasher, I realised it was one of my most important kitchen tools – the ease with which it deals with all the mess I can create in the kitchen is wonderful! Before owning a dishwasher I assumed it was intended only to wash up the dirty plates after a meal. Not so. Dishwashers can and do clean all the pans and tools used in the cooking – it is here they have their real impact on the cook. So make sure any dishwasher has a programme that is capable of dealing with the dirtiest cooking pans.

Types of Cooking Vessels

There is such a bewildering array of different pots and pans on the market to-day that it often seems impossible to decide which is best for any given purpose. Not only do cooking pans come in a variety of shapes and sizes, they also come made from a wide range of different materials. Here I will try to describe the advantages and disadvantages of the different materials that pans are made from and to suggest which shapes may be easier to use. However, as with most matters in the kitchen, what really counts is what you find works for you, so you will only really learn with a little experimentation and trial and error. Over the years, I have bought many pans and tins, but most have been discarded, or languish, long forgotten, at the back of a cupboard.

Frying Pans and Saucepans

Probably the most important consideration when looking at cooking pans is the material they are made from. Nearly all pans used on the hob are made from metals. Different metals conduct heat with differing efficiency. Copper conducts heat very well so providing a uniform temperature around the base of the pan even when the heat source is not itself heating evenly around the pan. Conversely, stainless steel is a very poor conductor of heat, so stainless pans can suffer from "hot-spots".

Most good quality stainless steel pans are made with a sandwich construction with an inner layer of copper and thin outer layers of stainless steel. Thus they combine the easy to clean properties of stainless steel with the good thermal conductivity of copper. While this makes for a good compromise (and is in fact the type of pan I favour myself) there can be disadvantages. For example, if the copper layer extends up the sides of the pan they can be heated to a very high temperature – the copper conducts heat from the hotplate inside the wall of the pan upwards. Where there is some water inside the pan it keeps the temperature of the stainless steel wall down to a temperature of no more than 100 °C. However, above the water line this cooling effect is lost and the temperature can rise – so that any food splashing above the liquid line can burn on the side of the pan.

Other commonly used metals include aluminium, which is nearly as good a conductor of heat as copper, and carbon steel and black (or cast) iron both of which are of intermediate thermal conductivity. Some years ago there was a trend for Pyrex glass saucepans; while Pyrex can withstand the intense heat of a hotplate, it is a very poor conductor of heat, so these pans while perhaps attractive to look at, are not a great deal of use for the serious cook.

The next important consideration is the thickness of the base and sides of the pans. Obviously, pans with thicker bases will tend to spread the heat from the hotplate more evenly, they will also take longer to heat up and will stay hot longer making very precise temperature control difficult when you want fast temperature changes. Generally, the thicker based pans will last much longer – pans with thinner bases often distort under the heat, or become dented through general wear and tear. Personally, I have one small pan with a thin base, while all the others I use have thick bases. I find this combination quite sufficient for all my needs.

The final important considerations are the shapes and sizes of the pans. I find a range of sizes of both frying pans and saucepans very useful. Small frying pans are particularly useful for such things as fried eggs and pancakes where you do not want the food to spread out too much while cooking.

Casseroles, etc.

When using a casserole dish in an oven, the actual material from which it is made is not as important as it is for saucepans. In the oven the outside of the casserole will be kept at the temperature of the oven while the inside will be kept at about 100 °C as the water inside gently boils. A material with a low thermal conductivity can be slightly preferred as it will withstand this temperature difference between the outside and inside more readily and will not lead to the rapid boiling away of the contents.

Indeed the most important feature of a casserole dish is that it prevents the liquid inside from evaporating too fast. The main way in which water is lost is from the escape of steam around the lid. So a well fitting lid is essential for a good casserole dish. In some dishes, the lids do not fit as well when they are hot as they do when they are cold. This can arise from the thermal expansion of the lid not matching exactly the expansion of the dish and is not something you can determine simply by looking at a dish. However, if the lid is heavy and flat, and the lip of the dish is also flat then the sheer weight of the lid should be enough to maintain a good seal.

Another consideration that suggests thick walled, heavy, casserole dishes are preferable is that they retain heat when taken out of the oven and so will continue the cooking while other ingredients are added, etc. Personally, I prefer cast iron casserole dishes, as these can be used both in the oven as well as on the hob; but many ceramic materials are equally useful.

Preventing foods sticking to the cooking vessels: Non-Stick Finishes

One of the major problems that many domestic cooks face is that their food sticks to the pan and burns. There are a number of different remedies. All work around the same general principle, the need to prevent chemical reactions between the food and the surface of the cooking vessel.

Proteins become chemically quite reactive at high temperatures (above about 80 °C) and will not only react with themselves to form networks (as in boiling an egg) but also with metal ions at the surface of cooking pans. You should be aware that there are metal ions in the glazes of ceramic pans so they can be as "sticky" as a metallic pan.

If the protein in your food does react with some metal at the surface, then it can literally become glued to the surface. Once a piece of food is stuck to the surface, then the temperature can rise above 100 °C if all the water is lost through evaporation and then the food can burn which can lead to unpleasant, bitter flavours.

So to prevent food sticking and burning, you need to prevent any protein molecules reacting with the surface of the pan. There are two techniques – placing a chemically inert layer between the food and the metal surface of the pan, and keeping the food moving so that no one part remains in contact with the same surface long enough to form a chemical bond.

In many cases simply stirring the dish while cooking is quite sufficient to prevent any problems. However, this is not always practical. For example, the cook may be called away to answer the telephone at a crucial moment! So you need some alternative strategies.

A useful approach, is to ensure that all the protein at a surface is already reacted before stopping stirring. For example, when cooking meats keep the pieces of meat moving until they are well browned. However the commonest solution is the non-stick finish.

Poly(tetrafluroethylene): teflon, etc.

Most non-stick pans have a coating of the inert polymer poly(tetrafluro-ethylene) or PTFE, known by such trade names as Teflon, etc. The PTFE layer will not react chemically with proteins and so prevents any sticking occurring. PTFE coated pans can be particularly useful in baking. When you cook a cake or bake bread, you simply put the dough in the baking tin and do not stir it around at all. There will be plenty of protein molecules in the bread dough or cake mix (from the flour and eggs – if any) that will stick to any exposed metal ions at the surface of the baking tin. Unless you prepare the tin very carefully and cover over all such metal ions with an impermeable layer, there is a real risk of the cake, or loaf sticking to the pan.

Do it yourself non-stick pans

As I have already noted, there always is a possibility of a cake sticking to the sides of the baking tin. Sticking is caused by the formation of chemical bonds between the protein molecules from the eggs and the metal surface of the tin. You can reduce the likelihood of any such bonds forming by preventing the contact of egg proteins with the metal of the tin. Greasing the tin with butter and putting some flour on the butter provides a barrier preventing the egg proteins from reaching the metal surface.

Some tins are less prone to sticking than others. The likelihood of sticking is reduced if there are fewer reactive sites in the metal surface; surprisingly if the metal is very clean then there are many reactive sites. The number of reactive sites may be reduced by reacting them with something else. This is often done by letting a 'patina' form by heating some oil in the pan until it is smoking. Normal washing up methods remove the patina so that the pans will stick again. You should never clean metal pans on which a patina has developed with any detergents or soaps, they should be wiped clean and left with a thin covering of oil. If the pans are ever cleaned with detergents then you will need to reform the patina from scratch. Many people are encouraged to learn that really good non stick cookware can be made from some of the cheapest iron or steel pans with a little preparation. It is even better when you realise they never need to be washed up! Not only do you save money, you also save house-work.

Ovens and Hobs

Probably the most important items of equipment in any kitchen are the sources of heat, the oven and the hob (I will not discuss microwave ovens here, they have been addressed already in Chapter 4). There are several different types of heating that can be used on hobs. The commonest are electric and gas rings, but there are also halogen and induction heating sources. Each has its own advantages and disadvantages.

Electric rings tend to warm up and cool down slowly which can make precise control of the heat entering a pan rather difficult – leading to the boiling over of pans and to burning of sauces, etc. However, when used in ceramic hobs they are remarkably easy to keep clean. Also, these days all households have mains electricity connected, but not everybody has a gas supply.

Gas rings are my personal favourite. The heat is immediately controllable simply by controlling the flow of gas. Halogen and induction cookers also provide instant control of the heat source but are more complex and so more prone to go wrong. Induction cookers can only be used with special heavy bottomed pans.

When it comes to ovens the range of choice is extensive – again the main sources of heat are electricity and gas (although oil and solid fuel fired ovens are available). There is a further variation in that many modern ovens incorporate a fan to circulate the hot air and maintain a uniform temperature inside. Such fan assisted ovens will normally heat up more quickly than conventional ovens with no fan, and will not have the variation of temperature (hot at the top and cool at the bottom) of conventional ovens.

Many chefs prefer the conventional oven as it has zones of differing temperatures allowing a wide range of different dishes to be cooked at the same time. In my own experience, the fan assisted ovens are probably easier to use, but are less versatile than the conventional form. When using any recipes, you should be aware that you need to set a lower oven temperature when using a fan assisted oven than when using a conventional oven. The oven manufacturers usually provide a conversion chart for this purpose.

However, setting the oven temperature is not enough by itself. I have taken a thermometer to many different ovens and found that there is a very wide difference between the temperature set on the dial and the actual temperature inside the oven. Most domestic ovens seem to have a temperature that is within 15 °C of the set temperature, but I have found ovens that are worse. The only solution is for you to get to know your own oven – if possible make measurements of the temperature inside the oven and draw up a chart so that you know what the temperature will actually be when you set a particular number on the dial.

I should note that all the recipes in this book give oven temperatures that are intended for use in a well calibrated conventional oven and should be adjusted to suit your own individual oven.

Some Experiments to try at home

Both these experiments require careful supervision by a responsible adult. They should only be performed out of doors over a small portable gas stove. There should be a bucket of water available close to hand so that if anything should catch fire it can be dropped immediately into the water. Also protective clothing including glasses should be worn to prevent any injury.

Cooking an egg on a piece of paper

This is a simple experiment to illustrate how cooking food will keep the temperature of the surface of the cooking vessel down to a temperature of boiling water (ca. 100 °C).

You will need a small camping gas stove, an A4 sized piece of clean white paper, a little cooking oil, an old metal coat hanger, a few large paper clips, a metal spatula and an egg. Also keep a washing up bowl full of water nearby in case of any accidents!

Begin by making your paper frying pan. Bend the coat hanger to form a rough square about 20 cm on each side with a handle you can use to hold it. Fold up the sides of the paper to make a square "dish" with a base to fit inside the square of the coat hanger and fix it inside with the paper clips. Spread a little oil on the paper (this will help to prevent the egg from sticking to the paper).

Next light the gas burner and set it on a low heat. Now break the egg and put it in the "frying pan". Now hold the frying pan a few centimetres above the flames of the gas burner. Make sure all parts of the paper that are near the flames have some of the egg covering them – **do not allow any parts of the paper that are not covered above with any egg to come into contact with the flames.** Keep moving the "frying pan" from side to side and up and down as the egg cooks. After a minute or two the egg should be fried and you can slide it from the pan onto a plate using the spatula. Do not eat this egg as there may be some contamination from charred paper.

If at any time the paper should catch fire immediately plunge the whole frying pan into the bowl of water to extinguish the fire and thus prevent any risk of injury.

The reason this works is that the egg white and yolk both contain lots of water. As the egg is heated, so this water heats up to 100 °C and then starts to turn to steam – while remaining at 100 °C. The fact that the paper is thin means that the temperature never gets significantly above 100 °C which is not hot enough to start it burning.

You will probably find that the paper becomes a little charred around the edges of the egg – this is the region where there is least water and where the temperature can rise high enough to start the paper oxidising. It is important that you keep the paper moving so that no such spots ever get hot enough to catch fire.

Boiling water in a balloon

This is a very similar experiment to the one described above. Instead of cooking an egg you boil some water in an otherwise inflammable thin walled container.

Again this experiment should only be performed under adult supervision and out of doors. You will need a small camping gas stove, a balloon, some water and a stand made from a few bent coat hangers to support the balloon.

Begin by making the stand for the balloon. Take 3 metal coat hangers and bend them so that they form a small tripod from which you can hang the balloon above the gas burner. Next partially fill the balloon with some water and tie it up with a piece of thread and hang it by the thread to the tripod. Now light the burner and leave on a low setting. Stand well back.

The water in the balloon will slowly be heated by the burner and will keep the rubber of the balloon cool enough that it neither melts or burns. As the water starts to boil so the balloon will start to inflate with the steam. Move the burner away from the tripod and turn the burner off immediately as the balloon could burst and spray hot water around. If the balloon should burst during the heating, do not panic, but turn off the gas burner using a oven glove in case the tap has got hot.

Meat and Poultry

Structure of Meat

It is perhaps, in the cooking of meats, that a good understanding of the underlying science can provide the greatest and most immediate improvement in one's own cooking; certainly this was true for me. Meats are prepared (boned, minced, ground, chopped, etc.) and cooked (grilled, roast, stewed, etc.) in so many different ways, that it is to me astonishing that all these diverse cooking processes involve the same few scientific principles.

Cooking meat is all about producing the right textures and flavours. Once you understand how cooking alters the texture of meats you will quickly be able to control the process so as to produce tender meats all the time. Equally, once you appreciate the complexity of the chemistry that goes into the development of the meaty flavours, you will soon learn that there are a few simple, but crucial steps that are essential to flavour development during the cooking process.

Before you can fully appreciate the changes that occur when meat is cooked, you need to have some knowledge of the structure and composition of meat before it is cooked. We are all familiar with the texture of meats; meat has a 'grain' being made up from muscles which are in turn made up from bundles of protein fibres. These proteins contract, when appropriate chemical signals are sent to them, so operating the muscles.

Between the fibres and the fibre bundles there is some 'connective tissue', which holds the muscles together and connects them to the bones. Muscles that have to bear greater loads, such as thigh muscles that operate the legs, etc., tend to have more of this connective tissue. All connective tissue needs to be very tough and strong (otherwise it would be no good at transferring the load from muscles to bones!). So muscles that have lots of connective tissue will make for tough and gristly meats.

There are three main types of connective tissue: collagen, reticulin and elastin. Collagen is the most abundant and the most important for the cook to appreciate. As is described in detail in Chapter 2, collagen is a complex molecule made up from three strands that are twisted together rather like a rope. Collagen derives its stiffness and strength from the arrangement of these intertwined helices. However, if collagen is heated to temperatures above about 60 °C, the three strands can separate and the material loses its strength. Once denatured into single strands collagen becomes a very soft material, and is given a different name, gelatin. You already know that gelatin is a soft, tender, material, since it is used as the basis of all jellies.

The collagen is mostly found around bundles of muscle fibres and helps to hold them together. The muscles are then joined to the bones with yet more sinews (yet more "connective tissue"), which cooks recognise as tough 'gristly' material. These sinews are made from the proteins reticulin and elastin; reticulin and elastin can only be denatured and softened by heating for very long times at temperatures above 90 °C.

What Makes Meat Tough or Tender?

Meat consists of muscle fibres, connective tissues and fats. The muscle fibres largely consist of two proteins, myosin and actin. In a living animal, these proteins are able to change their shape reversibly so that the muscle contracts in response to a chemical signal. You can think of a muscle fibre being rather like a piston in a cylinder. If air is sucked out of the cylinder the piston moves in and the whole thing contracts, when air is allowed back in the piston moves back out. In muscle fibres the proteins are extended along the fibres and can, to some extent move past one another, to cause a similar contraction, thus moving the bones to which they are attached.

When muscle fibres are heated above about 40 °C the proteins start to denature, that is they change their shape irreversibly (see Chapter 2 for a detailed description). In muscle proteins, which are extended along the muscle fibres, this change of shape involves the proteins coiling up. This coiling process inevitably causes some contraction of the muscle. As the muscles contract in cooking, so pieces of meat contract along the direction of the muscle fibres – something we have all seen when cooking meats. When the protein molecules are denatured and the muscles shrink so the meat becomes harder. If you tense your muscles you can easily feel the same thing – as the muscles contract, so they become quite hard (just how hard will depend on how fit you are!). In cooking terms we would say that when muscles are contracted like this the meat will be tough.

Now you can see that heating meat will tend to make it tougher. As meat is cooked, so heat flows in and more proteins are denatured. The denatured proteins shrink making the meat progressively tougher. The longer you cook any meat the 'tougher' the muscle fibres will become.

Figure 6.1. A series of sketches illustrating the structure of muscles. At the top is a muscle which is made up from bundles of muscle fibres (shown below the whole muscle). Below the bundle of muscle fibres a single fibre is sketched and at the bottom there is a representation of the actin and myosin proteins that slide past one another to make the fibres and hence the whole muscle contract

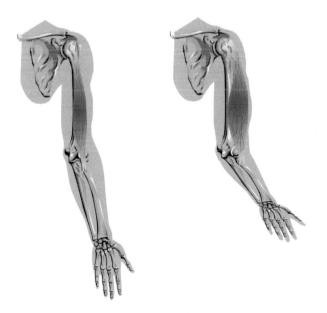

Figure 6.2. Muscles contract in the body to move bones

On the other hand, the connective tissues that join the muscles to the bone and wrap around the muscle fibre bundles, are too tough for us to bite through (and remain largely indigestible) before they are heated. However, after prolonged heating to a temperature above 60 °C the collagen triple helices are destroyed and the tough collagen becomes the soft gelatin. Accordingly there must be a compromise between overheating the muscle fibres and producing a tough product, and not heating enough to denature the collagen which would again leave tough meat.

Another important component in meats is the fat. Animal fats are solid in uncooked meats but melt during cooking and will often run out into the cooking pan. These fats provide a good deal of the flavour of meats – as we have seen in Chapter 3, flavour comes from small molecules and in uncooked meats the only small molecules are the fats. The fats also act as a sort of lubricant, so that if there is some fat close to a toughened muscle fibre it can act to make it appear less tough when we eat. For fats to act in this way it is necessary that they are intimately mixed into the meat (i.e. marbled through the meat) and not well separated from the muscles as is often the case. This is why steaks "marbled" with fat are more highly prized than those with only a little fat on the outside; of course, from a health viewpoint, the marbled steak may be less desirable as it will have a larger fat content.

The remaining component of meat is water. In fact most of meat (about 60%) is water, making it the largest single component of meat. You might think that this water is not very important in cooking, but it does play several different roles that help determine both taste and texture. When meat is cut some water flows out and the meat becomes a little wet. However, most of the water in a piece of meat remains locked in (technically referred to as "bound water"). The water molecules are actually trapped by some of the proteins in the meat. When the proteins are denatured by heating some of this 'bound water' can escape. You can see this when you fry meat; after a short cooking time, some liquid (mostly water) will start to flow out of the meat and as it boils in the hot fat it 'spits' at you. If significant amounts of water are lost in this way the cooked meat will seem to be rather dry.

The Composition of Different Meats

	% Water	% Protein	% Fat
Beef	60	18	22
Pork	42	12	45
Lamb	56	16	28
Turkey	58	20	20
Chicken	65	30	5

If meat is frozen before it is cooked the act of turning the water into ice crystals can release it from its surrounding proteins so that when the meat is thawed it can flow away much more easily. This freeing of bound water on freezing is why frozen meats often seem to be drier than fresh ones.

What Gives Meat its Colour?

Many people think that the red colour of meats comes from the blood in the veins and arteries. Of course, a little thought will demonstrate that matters cannot be that simple. When an animal is slaughtered, the blood drains out of the veins and arteries so the colour must come from elsewhere. The red colour of the blood comes from a special molecule, haemoglobin, which is used to carry oxygen around our bodies in our blood. Some of this haemoglobin will, of course be absorbed into the muscles and will thus contribute to the red colour of meat. However, there is not enough haemoglobin in muscles to make meat as red as it is.

The chemical process that makes the myosin and actin molecules in muscle fibres slide past one another uses up oxygen. However, if the muscle is to do a lot of work, it may not be able to get all the oxygen it needs from the blood. Instead some oxygen is stored in special proteins (myoglobin) in the muscles. These myoglobin molecules are similar to the haemoglobin in the blood and like haemoglobin they also have a red colour when oxygenated, and adopt a purple hue when they have given up their oxygen.

Animals need different amounts of myoglobin in different muscles. In general the amount of myoglobin is related to how much a muscle is used. Heavily used muscles need lots of myoglobin, and hence are dark, while infrequently used muscles need little myoglobin and the meat from these muscles is light in colour. Thus in turkeys which stand around a lot, but hardly ever fly, the leg meat is dark, while the breast is white.

Similarly, game birds which do fly a lot have dark breasts. In fact all game tends to consist of dark meat since wild animals tend to use all their muscles, while domesticated animals seldom need to.

There is a further distinction between red and white meats. There are two types of the protein myosin which can be found in muscle. These different types (which are found in different types of muscle fibre) use different chemical routes to provide their energy.

The 'slow' fibres burn fats to provide the energy; these muscles need oxygen to operate. The 'fast' fibres burn glycogen, and do not need to use any oxygen. Accordingly, muscles made from fast fibres do not need any myoglobin and are always white. In practice, 'fast' fibres cannot operate for long periods. So the proportion of 'fast' and 'slow' fibres will depend on the use a muscle is given. For example, we all use our leg and back muscles all the time, just to stay upright, so these muscles need to contain almost entirely 'slow' fibres. In contrast, fish are supported by the water in which they swim and do not need to use their muscles to stay still, accordingly they can have muscles that are predominantly made from 'fast' fibres. This difference in type of muscle fibres underlies the different colours of mammalian and fish flesh.

Cooking Meat – Tough or Tender?

Different muscles in different parts of an animal (for example in a turkey the breast, legs, wings etc.) have different ratios of muscle fibres, connective tissues and fats; and the types of muscle fibres may also be different according to the amount of use a particular muscle has had. On heating, the muscle proteins start to denature at temperatures above 40 °C, and then coagulate into hard knotty lumps at temperatures above about 50 °C. Both these processes make the muscle fibre part of the meat tougher. However, it is not until the temperature is even higher (above ca. 60 °C that the first of the connective tissue, the collagen, starts to soften and turn into gelatin, so making 'gristly' meat more tender. Thus, cooking meat will always be a compromise between preserving the tenderness of the muscle proteins and softening the connective tissue.

The general rule of thumb is to cook meats with low content of connective tissue for a short time; and to cook meats with a lot of connective tissue for a long time.

For example, in a turkey there is more collagen in the wings and legs than on the breast; thus to produce a uniformly cooked bird we should try to cook the breast less well than the other parts. This means that the temperature in the breast meat should be lower than that in the legs and wings. We can protect the breast from overcooking by partially insulating it from the heat of the oven, for example by covering it with some paper or foil.

Slaughter and Ageing

The way in which animals are slaughtered, and how their meat is stored after death can significantly affect the texture and flavour of meat.

Just because an animal is dead it doesn't mean that all the various chemical reactions that take place all the time when the animal was alive stop immediately. In fact, lots of chemistry continues in the muscles, leading to rigor mortis. After slaughter, the muscles continue to work, using up the oxygen that was stored in the myoglobin. Since the blood is no longer flowing, the waste products are not carried away, and accumulate in the muscles. The most important of these waste products is lactic acid. We are all familiar with the feeling of excess lactic acid in our own muscles. When we take unfamiliar exercise we feel 'stiff' the following day. This stiffness is due to an excess of lactic acid in the muscles that were worked more than usual and had to use up all the oxygen in the myoglobin just to keep going. Since we made so much lactic acid our blood could not carry it all away and it started to attack our muscles leading to the stiff feeling the following day.

The state of the muscles immediately prior to slaughter is important, since it determines how the chemical reactions occur after death. If the muscles are tensed, or have been unusually heavily used, just before slaughter, then they will have used up a lot of the available oxygen. Since the animal was still alive it would have had a working circulatory system and most of the lactic acid formed would have been removed in the blood. In the animal after slaughter, there is likely to be a deficiency of oxygen in the myoglobin, so that little, if any, further lactic acid is produced and little breakdown of the proteins occurs on ageing. The result is that the meat will be tough and not very flavoursome. Conversely, if the animal is relaxed before slaughter, then there will still be sufficient oxygen in the myoglobin to allow the biochemical reactions to continue to make lactic acid even after death. This lactic acid will not be carried away by the blood (as the circulation stops after death) and can build up in the muscles and begins to break down the muscle proteins and the connective tissue. The result will be meat that is more tender and more flavoursome.

Lactic acid is very important since it makes the environment around the muscle fibres quite strongly acid and starts to denature the proteins and connective tissue. The proteins are not only denatured but can also be broken down into smaller molecules which promotes flavour. Any degradation of the connective tissues will, of course make the meat more tender, so the production of this lactic acid serves to increase and change flavours and to tenderise meat. Since the high acidity of the meat also inhibits the growth of bacteria, the meat can be stored, or aged, before use.

The ageing of meat alters, and most say improves, the flavour. The changes in flavour are caused not only by the action of the lactic acid but also by enzymes which help to break down the proteins into smaller flavour molecules.

Cooking Meat – How much Flavour?

Flavour is another major consideration when deciding how to cook meat. When proteins are heated together with sugars (nb in this context 'sugars' include large molecules made by joining small sugar rings together such as polysaccharides or starch and other carbohydrates – see Chapter 2 for details) to temperatures above about 140 °C, a whole series of chemical reactions occur (the Maillard reactions – see Chapter 3 for a more detailed description). These reactions (which also brown the meat) break the large protein molecules down into smaller molecules that are volatile and hence can release flavours and smells. The flavours we think of as 'meaty' smells in fact are only generated during cooking at these high temperatures.

Thus, if we do not allow the meat to get hot, none of these Maillard reactions will occur and the final dish will not taste very 'meaty'. To achieve the desirable 'meaty' flavours we need to make sure that some parts of meat reach high temperatures (well above 100 °C) and remain at those temperatures for long enough for the meat to become a rich dark brown colour.

The combination of attempting not to heat those muscles that contain little connective tissue above about 40 °C, while heating those parts where there is lots of connective tissue to temperatures above 70 °C; and at the same time ensuring that some parts are heated to above 130 °C makes the cooking of meats a complex process.

However, there are a few simple guidelines that, if followed, should ensure a good, tender, and flavoursome result every time.

Key Points to bear in mind when cooking meats

- Always ensure the outside of the meat is cooked at a high temperature until it is a dark brown colour.
  - i.e. start cooking meats at high temperatures
- Cook meats with little connective tissue for only a short time so that the outside is browned, but the inside does not become tough
  - i.e. grill, fry or roast these meats.
- Cook meats with lots of connective tissue for very long times so that all the connective tissue denatures and the bundles of coagulated muscle proteins fall apart making the meat tender again.
  - i.e. make stews with gristly meats

These key points are illustrated in the recipes below, for steaks, roast leg of lamb and for a rich beef stew.

Why are these key points important?

Cooking at high temperatures and browning the meat is most important since the main flavour is generated by the "Maillard" reactions which only occur at temperatures above about 130 °C. The cooking time should be adjusted to be just long enough to degrade the connective tissue present in the meat, without toughening the muscle proteins too much.

The Effects of Heat on Meat

Temperature	Colour	Muscle Proteins	Protein bound water	Collagen
40 °C	red	denature		
50 °C		start to coil up and shrink	begins to flow	
60 °C	pink	coagulation well under way		begins to denature
70 °C	grey	mostly coagulated	flow ceasing	
80 °C	light brown	densely associated tough meat		
90 °C				rapidly turning into gelatin
100 °C				

Roast Disasters

One of my relatives – if I said which one my life wouldn't be worth living, but maybe she will recognise herself anyway – would not claim to be one of the world's great cooks, nor would she admit to being amongst the worst. But on one memorable occasion when several family members and their partners were visiting she managed to produce probably the most ghastly roast of all time. We had been kept waiting for quite some time for dinner, not unusual as kitchen organisation is not one of her skills either, but eventually plates with roast potatoes, some green vegetables and some slices of a roast meat were put in front of us. The meat was of a uniform grey colour, with a coarse texture. I certainly could not identify the meat at all from its appearance. When our hostess left to go out to the kitchen, I quickly asked the others at the table what it was we were eating – no-body knew – pork, beef, lamb and even turkey were offered as possibilities, but the texture and colour were not like anything any of us had experienced before.

The meat had been cooked for a very long time – far too long, and at a low temperature (to prevent the oven getting dirty!) so that it was just a dirty grey colour with no significant browning. At the same time the long cooking had coagulated the muscle proteins to give a particularly coarse texture – more usual in stews that have been simmered for many hours. To this day, all those of us who were present at this family meal hold regular discussions as to the identity of the mystery joint – nobody has ever plucked up the courage to ask the cook exactly what she had prepared, and I doubt that she would even remember now, so this will permanently remain one of life's great mysteries for me.

Cooking Steaks

We begin the recipe section with a few notes on cooking steaks. In 'science speak' steaks should always be selected from muscles that contain the minimum amounts of connective tissue, or in 'cook speak', use tender cuts! Cooking of steaks is all about getting enough chemistry to go on at the surface to provide flavour, without causing too much denaturing of the muscle proteins inside, which would lead to tough meat.

Key points to remember are:
- Remove as much connective tissue as possible before cooking
- Cook at high temperature for a short time
- Make sure the surfaces of the steaks are well browned

Sautéed Sirloin Steaks

Ingredients (for 6)
1 kg Beef Sirloin
50 ml good quality oil
Fresh ground Pepper

Method

Begin by preparing your steaks. Cut the sirloin into slices about 2 to 3 cm thick and then divide into 6 steaks (alternatively get your butcher to do this for you). Next trim away any connective tissue you can see, remember you will only be cooking the steaks for a short time and this connective tissue will be tough and inedible. Grind a little pepper over the steaks. Some people recommend grinding a little salt over the steaks as well, but I prefer not to use salt – try both for yourself and find your own preferred taste.

Heat the oil in a thick based frying pan until it is almost smoking. Add the steaks, a couple at a time to the pan (don't try to cook too many steaks at one time since putting cool steaks in the pan lowers the temperature). Make sure the steaks are touching the bottom of the pan so that they are heated as quickly as possible, and at the highest temperature possible. Cook for about 30 seconds and turn the steaks over, cook for another 30 seconds and turn over again. The initial searing you give the steaks will help to prevent any sticking. Keep on cooking for about 2 minutes longer until the bottom is a deep brown colour and then turn again and cook that side for another 2 minutes. Keep the steaks warm and cook the rest in the same way. You will find that the steaks you cook last will brown most quickly.

I have asked several food scientists why this happens and have been given two different explanations. One suggestion is that the Maillard reactions that cause the browning and give the flavour, may, in chemical speak be "autocatalytic". This means that once some of the "reaction products" (in cook speak – "brown bits") have accumulated in the "reaction vessel" (frying pan to real people!), the reaction will go faster. In plain language, browning takes place faster in a pan in which some meat has already been cooked. The alternative explanation is that some of the browned meat breaks off from the pieces cooked first and remains in the fat coating the meat cooked later.

Whatever the underlying cause, you can take advantage of the difference in browning times to serve your guests steaks according to their tastes. If the steaks are cooked in three batches and kept in a warm oven (at ca. 50 °C) before being served, then the first cooked steaks will end up well done, the second pair will be medium and the last which are served straight from the pan will be nice and rare. If you have sensible guests who all want rare steaks, don't panic, there is an easy solution. Simply cut off a small piece of meat from the sirloin (preferably one that has the most gristle or connective tissue) and begin by cooking this at a very high temperature for about five minutes until it is almost burnt, so that the pan has plenty of 'brown bits' or 'reaction products' to catalyse the Maillard reactions for the steaks. Now you will be able to cook the 6 remaining steaks in about 3 minutes each, so they all end up nice and rare.

What could go wrong when cooking steaks and what to do about it

Problem	Cause	Solution
The steaks are tough	Steaks cooked too long	Next time don't cook them as long. For now use a sharp 'steak knife'!
	Poor quality meat (e.g. animal stressed before slaughter)	Use a different butcher to buy your meat. Look for purple, rather than red meat (see the panel on slaughter and ageing) or use a "better" cut.
Poor Flavour	Outside not well browned	Cook for longer and/or cook at a higher temperature
Too rare for your taste	Outside well browned, but inside still purple	You could cook the steaks for a longer time at a lower temperature so that the middle gets pink – or develop a taste for rare meat! Or use thinner steaks. For now simply cook for a little longer.
Too well done	It took so long to brown the outside that the inside was overcooked	Cook at a higher temperature, or use the "autocatalytic" effect by cooking a little meat in the pan first. Or use thicker steaks.

Recipe Variations

There are endless variations you can produce on the basic steak recipe. You can change the cut, the best cuts of beef for steaks are Fillet, Sirloin, Rump, and T-bone. You can also cook lamb or pork chops in the same way, or take steaks from venison and other game.

Grilled Steaks

Grilling steaks differs in that you do not cook the steaks in their own juices as in sautéing, so there is no autocatalytic effect to accelerate the browning (or Maillard reactions). The recipe is exactly as above except you should brush a little oil or melted butter on the steaks before putting them on the grill; the oil will help to make sure the heat is spread evenly over the surface of the meat. The cooking times are about the same, but will depend on the power of your grill, so a little experimentation is called for.

Steaks with a mushroom and brandy sauce
There are many recipes for steaks that include a simple sauce made in the pan using the juices that have come from the steaks during cooking, this is a very simple recipe.

Ingredients (for 6)
1 kg Beef Sirloin
50 ml good quality oil
Fresh ground Pepper
200 g sliced mushrooms
200 ml stock (either from a stock tablet or prepared as described in Chapter 9 or in the stew recipe below)
20 ml brandy
100 ml cream

Method

Begin by preparing the steaks exactly as in the basic sautéed steak recipe above, but cook for a little less time as they will be flambéed later. Keep the steaks warm – either in the oven or covered with aluminium foil. Add the mushrooms to the frying pan in which the steaks were sautéed and cook on a medium heat until they become soft (about 2 to 3 minutes) stirring all the time. NB the mushrooms will pick up a lot of brown colour from the juices in the pan. Add about half the stock and scrape all around the pan to get all the juices and brown bits that are sticking to the pan into the sauce which will turn a beautiful deep brown. Add the remaining stock and, if you wish, thicken using a teaspoon of cornflour which has been suspended in a little cold water (see Chapter 9 for details of how this thickening method works). Allow the sauce to come to the boil and then put the steaks back in and add the brandy (which you have warmed in a small glass). Immediately set fire to the alcohol fumes that come off the pan. You can either do this with a match, or, if you are using a gas stove, just tip the pan so that the gas flames up around the side of the pan to ignite the brandy fumes. Shake the pan from side to side until the flames die away, stir in the cream and serve immediately.

Garlic Roast Leg of Lamb

This roast is one of my favourites, it combines the delicate flavour of lamb with the richness of a gravy based on roast garlic. Don't worry that you will suffer from garlic breath after eating this dish, even though it uses a lot of garlic the roasting seems to alter the penetrating qualities of the garlic and it does not hang around for long at all.

Ingredients (for 6)
Leg of lamb (or half a leg – preferably the knuckle half, depending
on how greedy you are)
3 – 6 cloves of garlic (or more – depending on how much you like garlic!)
50g butter
1 litre stock (either from stock tablets or prepared as described
in Chapter 9 or in the stew recipe below)

Method

Begin by preparing the lamb. Wipe the surface of the joint with paper towels so that it is dry, with a sharp knife score a few cuts about 5 mm deep into the joint. Next prepare the garlic butter – crush all the garlic and beat into the butter. Put the leg of lamb into a roasting pan – the pan should be just a little larger than the leg so that it can lie down comfortably on the bottom of the pan. Spread about two thirds of the garlic butter on top of the joint and the remainder on the cut end, working the butter into the cuts you made earlier. Heat the stock to boiling and pour about 200ml into the pan. Put the pan into a preheated oven at 180°C and cook for about 75 minutes.

After about 40 minutes the stock should have all evaporated, do not add any more stock but rather allow the juices and melted butter on the bottom of the pan to brown off.

After 75 minutes take the joint out of the oven and knock most of the roast garlic off the lamb and into the pan. Take the lamb out and set aside on a carving dish. Add about half the stock to the pan and scrape all the browned juices and other material into the stock – pour into a saucepan and then add the rest of the stock to the roasting tin and repeat to make sure you have removed everything out of the roasting pan you can possibly manage – remember it is all this lovely browned material that has all the flavour you want in the gravy.

Now you can bring the gravy to the boil and add pepper (and salt) to your taste. Finally you can thicken the gravy. You can do this either by adding about a teaspoon of cornflour suspended in cold water and stirring that in, or, to make an even richer gravy, by preparing a dark roux with about 40 g flour and 50 g butter in the roasting tin and then adding the unthickened gravy a little at a time to the roux making sure to stir all the time. The advantage of the first method is that there is no risk of getting any 'lumps in the gravy', on the other hand the second method does make a richer product. Whichever method you use you can find details of why it works and what can possibly go wrong, etc. in Chapter 9.

Finally carve the meat and serve with roast potatoes, green vegetables and the gravy.

Key points
- Don't add too much stock at the beginning – let it all evaporate in the first 40 minutes of the cooking time and then brown off to give lots of flavour
- Make sure you scrape out (deglaze) the roasting pan very thoroughly.
- Avoid cooking too long – leave the lamb 'pink'

What might go wrong and what to do

Problem	Cause	Solution
Under cooked – lamb 'still purple when carved'	Either the oven temperature was too low or the cooking time was too short	Next time cook at higher temperature or for longer times. For now carve the meat thinly and either cook it through using a microwave oven, or quickly sauté in a very hot, heavy based pan
Meat over cooked and tough	Either the oven temperature was too high, or the cooking time was too long	Next time cook for a shorter time or at a lower temperature. For now all you can do is carve very thin slices so that the meat is easier to chew.
Gravy has a pale colour or lacks flavour	Stock didn't evaporate during cooking	Next time use less stock to start with, or use a larger roasting dish so the stock is less deep. For now you can boil off the remaining liquid in the stock on top of the stove and allow the garlic, etc. to brown before making the gravy. Alternatively you can prepare a dark roux (see Chapter 9) to thicken the gravy
	Garlic didn't roast well ('it didn't get a deep colour)	Next time use a hotter oven. For now, you can roast the garlic in a thick bottomed pan on top of the stove on a very high heat.

Cooking times in minutes for various roast meats based on the size, rather than the weight of the joint – for an oven temperature of 180 °C

Shortest distance across joint (cm)	Beef (rare) (minutes)	Beef and Lamb (well done) and Pork (minutes)	Lamb (pink) (minutes)	Poultry (stuffed) (minutes)
5	25	30	25	35
7.5	50	75	60	80
10	90	130	105	140
12.5	140	210	170	230
15	200	300	240	320

Recipe Variations

Roast Turkey

Roasting poultry can be challenging as the different muscle groups on the breast, wings and legs, will generally benefit from different cooking times and temperatures. The breast meat in domestic poultry will generally contain very little connective tissue (since domestic birds do not generally use these muscles to fly!) and will benefit from cooking at a high temperature to bring out the flavour through the Maillard reactions, for only a short time so preventing significant coagulation of the tender muscle fibres. Conversely, the muscles in the legs and wings generally have a high connective tissue content and will benefit from cooking for longer times at lower temperatures. Accordingly, the cooking of a turkey will always be a compromise between the demands of the different parts.

There are several strategies that you can adopt to assist.

The best method is that used by professional chefs. They simply split the bird into separate pieces and cook the breast, legs and wings separately at different temperatures and for different times. They will then bring out a whole bird that has been cooked and looks nice and browned on the outside to show the diners, then take it back to the kitchen "to be carved"; but actually will carve meat from the separately cooked joints. The whole bird can then be kept ready to be brought out for the next customers to see!

However, for the domestic cook this strategy is not always acceptable so alternatives are needed. For example, you can protect the breast by covering it with aluminium foil. The idea is to keep the breast cooler than the legs and wings and so try to reduce the degree to which the muscle proteins coagulate, while at the same time allowing the collagen in the legs and wings to be heated to a sufficient temperature that it will become denatured.

Another technique that is often employed is to cook the bird over a pan of stock – the stock boils and keeps the bird moist as well as limiting the temperature to that of boiling water (100 °C). This method has the advantage that the cooking is quite slow and the collagen in the legs and wings can be denatured without burning the outside. In the recipe below, both techniques are combined. In the first stage of cooking, the breast is covered with aluminium foil which insulates it from the steam from the stock – during this stage the legs and wings are heated by the steam from the stock and the connective tissue is denatured making the meat tender. The second stage starts when the stock has evaporated. Once there is no water, the actual cooking temperature will rise. When water was evaporating the cooking temperature was limited to 100 °C even though the oven was set at a higher temperature. The steam rising from the water and surrounding the bird will have remained at 100 °C, or slightly higher. Once all the water has evaporated, the aluminium protection is removed from the breast and, as the temperature around the bird rises the Maillard reactions begin on the surface of the turkey. With a little practice it is possible to produce a turkey where all parts are cooked almost to perfection.

Ingredients
Turkey
Butter
Stock (prepared from the giblets)

For the Chestnut Stuffing
400 g chestnuts (or 250 g chestnut puree)
150 ml milk
100 g breadcrumbs
100 g bacon, cut into small pieces and fried
20 g fresh parsley, finely chopped
salt and pepper to taste

Method

Begin by cleaning the turkey and preparing the stock. Chop up and brown the giblets and add to a good vegetable stock (see Chapter 9), simmer for at least half an hour. Wash out the body cavity of the turkey with running water until all the blood has been washed away. Wash the outside skin in more clean water. Dry the bird on both the inside and outside using paper towels.

Next prepare the chestnut stuffing. Peel the chestnuts (see panel for a simple method) and simmer them for about 20 minutes in the milk. Make a puree of the milk and chestnuts and add the breadcrumbs, the fried bacon pieces and the herbs and seasoning. Mould the mixture into small balls ready to stuff into the turkey.

Put the chestnut stuffing on the bottom of the body cavity of the turkey leaving about 1cm of clear space above the stuffing and below the rib cage to allow hot air to circulate inside the bird as it cooks.

Put the turkey in a roasting tin and pour about 200 ml of the stock into the bottom of the tin. Spread some butter over the breasts of the turkey and then cover the breasts with aluminium foil – try to tuck the foil in so that it covers the breasts and not the legs, or wings. You may find it helpful to use a skewer to fix the foil – if so use a wooden, rather than metal skewer since a metal skewer would conduct a lot of heat into the breast and could lead to over cooking.

Cook the turkey in an oven at about 170 °C to 180 °C; a guide to the cooking times for turkeys is given in the table below. Check regularly to see that the stock has not all boiled away and top up if needed. About 30 to 40 minutes before the end of cooking remove the foil, drain off any remaining stock and allow the temperature to rise and the turkey to brown well. After 30 minutes test to see that the turkey is fully cooked by piercing the leg with a sharp knife and looking at the juices that flow out. If the turkey is properly cooked these juices will be clear. If they are pale pink, the turkey needs to be cooked for longer. Once the turkey is fully cooked and well browned on the outside, remove from the oven and allow to cool for an hour or so before carving. This cooling will allow some of the

Cooking times for turkeys based on the weight of the turkey

Weight of turkey (Kg)	Cooking time in a dry oven (i.e. without using stock) at 180 °C (or 160 °C in a fan oven)	Cooking time in stock at 180 °C (or 160 °C in a fan oven)
4	3 hours and 15 minutes	2 hours and 15 minutes
4.5	3 hours and 30 minutes	2 hours and 25 minutes
5	3 hours and 40 minutes	2 hours and 40 minutes
5.5	4 hours	2 hours and 50 minutes
6	4 hours and 15 minutes	2 hours and 55 minutes
7	4 hours and 40 minutes	3 hours and 15 minutes
8	5 hours and 15 minutes	3 hours and 40 minutes

gelatin made by denaturing the collagen in the connective tissue to set which will make the meat much firmer and much easier to carve.

N.b. the cooking times when cooking in stock are shorter than those in a dry oven because the steam from the stock actually improves the heat transfer from the oven heater to the bird so making it heat up more rapidly

---

Peeling Chestnuts

Peeling chestnuts can be difficult, the nuts have an inner papery skin which is held tightly in the convolutions of the kernel of the nut and an outer, tough rubbery skin which can be difficult to cut or tear. The traditional method to peel chestnuts is to boil them for a few minutes which makes the outer skins soft enough to cut and then using your hands tear off the outer skin and prize the inner skin away from the nut. While this method certainly will work, it will, as I can testify, lead to many a burnt finger. So here is a simpler, and safer, way to peel chestnuts that relies on the appliance of a little science. As described in Chapter 4, microwaves heat water – now in chestnuts there is little water in either the outer skin, or the papery inner skin, so microwaves will penetrate through the skin and heat the water in the nut itself. As this water is heated so some steam will be generated and this will have to go somewhere – it will "blow off" the papery inner skin. As a little heat penetrates into the outer skin so it will become softer and "tearable". So, if you have a microwave oven you can easily and painlessly peel your chestnuts. Simply pierce the ends of the chestnuts and put them about 10 at a time in the microwave oven and cook on high power for about one minute. Note that the piercing is very important – if you do not pierce the skin of the chestnuts there is a real risk that they will explode in the microwave oven. The amount of the steam generated inside the nuts can be quite large, so if there is nowhere for the steam to escape the pressure inside the chestnuts will build up until they explode! Once you get the conditions just right (by adjusting the cooking time and number of chestnuts used to suit your own microwave oven), when the chestnuts come out of the oven you should be able simply to squeeze them and the nut will pop cleanly out.

Basic Rich Beef Stew

This simple dish illustrates several different principles. How to use the Maillard, browning, reactions to produce flavour, how to take tough meat with lots of connective tissue and cook it so that it becomes tender, and how to use stocks to provide richness and thickness to a stew. There are many variations of this basic stew including, oxtail soup, goulashes, boeuf bourguignonne, etc. that are described at the end of the chapter.

Ingredients (for 4 people)
1.5 litres of good quality stock made from
250 g carrot   (one large carrot)
250 g leek   (one leek)
250 g parsnip (one medium sized parsnip)
300 g onions   (two medium onions)
250 g potatoes (two medium potatoes)
200 g mushrooms
2 litres water
(see Chapter 9 for a detailed description of stocks)

For the Stew:
500 g stewing beef (nb you should not use high quality lean meat, rather use a cut with plenty of connective tissue, etc.)
250 g Carrots
250 g Onions
1 – 2 clove garlic
20 g   Fresh parsley
20 g   Fresh rosemary

8 dumplings
150 g beef suet
300 g self raising flour
about 20 ml water

Method

The stock

Begin by making the stock. This will take about an hour and can be done well in advance. If you wish to use a ready prepared stock, or stock tablets, these will work fine, but will lead to a less rich flavour in the final stew and you will need to thicken the stew more at the end.

Clean and chop the vegetables into small (about 1cm) sized pieces. Put them in a thick bottomed enamel or stainless steel pan on a medium low heat. Do not add any water or fat or oil. Put the onions and leeks in first followed by the other ingredients finishing with the mushrooms. Keep the pan covered tightly and stir the vegetables regularly until they have softened and collapsed down to about half their original volume. Now keep a careful eye on the vegetables and let them begin to brown on the bottom of the pan. Once a layer of browned vegetable matter starts to stick to the bottom of the pan turn up the heat and let the colour deepen to a chocolate colour and then add a little boiling water. With a wooden spatula, scrape the browned vegetables off the bottom of the pan and add more water as needed. Add more water until it covers the vegetables and then add about half as much again – this should be about 2 litres of water. Continue to cook the covered vegetables on a low heat for another 40 minutes and then strain and scrape the mixture through a sieve. Keep the liquid that has been strained as your stock for the stew.

Browning the meat – developing the flavour through the Maillard reactions

Cut the meat into pieces about 1 – 2 cms in size making sure that the pieces are no more than 1cm thick in the direction of the grain of the meat. Heat some fat (preferably beef dripping, but any fat or oil will do) in a thick bottomed frying pan. You should have enough fat to cover the pan in a layer about 1mm thick (about 20g should suffice). When the fat is very hot add the pieces of meat, a few at a time so that they form a single layer in the pan. Keep stirring the meat until it is well browned. You should be aiming to get the surfaces of all the pieces of meat to be dark brown colour (with an almost polished look) like an old mahogany table, or freshly ground roasted coffee. (If you have a smoke detector in your kitchen it may be a good idea to take the battery out – I always set my detector off when I cook meat in this way!). Once the meat is browned put it in a casserole dish, if necessary repeat until all the meat is browned.

The next stage is to "deglaze" the pan – that is to collect all the flavour that has been developed during the browning of the meat and make sure it all ends up in the stew. Once all the meat has been cooked turn down the heat and add a little of the stock to the frying pan and scrape around the pan to collect all the brown, 'burnt on bits' that have collected in the pan. It is most important for the flavour of the dish that you make sure you scrape the pan out very thoroughly. Pour this stock into the casserole. Repeat several times until the stock you add to the frying pan gets no darker when you scrape around.

Stewing the meat – tenderising by denaturing the connective tissue
Now add the rest of the stock to the casserole dish which should be less than half full at this stage, put on a tight fitting lid, and cook the stew either on top of the stove on a very low heat or in the oven at about 160°C for about 3 to 4 hours, longer for tougher meats. You will need to check every half hour or so that the casserole is not running dry as water evaporates; if the liquid level falls add a little water, or stock, to keep all the pieces of meat below the level of the liquid. You can do all this cooking in advance and you can break the cooking overnight if you wish. I usually make the stock and cook the stew for about an hour on one day and then put the casserole in the fridge overnight before finishing off the cooking the next day.

After 3 hours cooking you should test that the meat has become tender. There are two methods. Either take a piece of meat out of the stew and try eating it, or use a blunt knife to see if it can cut through the meat easily. If the meat is not tender enough keep on cooking until it is! You should avoid cooking the meat for much more than 4 or 5 hours as it will fall to pieces and become rather mushy – the best course of action is to test the meat regularly and often.

Finishing the stew
The next stage is to make sure you are happy with the flavour of the stew. You really need to taste very carefully and decide exactly what you want to add. My suggestion is to add a crushed clove of garlic, a medium sized sliced onion (which has been sautéed until it is just starting to brown), a large sliced carrot, a teaspoon of fresh chopped herbs (preferably parsley and rosemary – but the choice is yours) and some fresh ground pepper and salt to taste. You might also add a little port and redcurrant jelly to add richness. The best way to set about adjusting the flavour is to add about half the garlic and onions and then stir for a few minutes before tasting. Now add other ingredients a little at a time until you are completely satisfied with the flavour. It is probably best the first time you cook this recipe to follow the ingredients and amounts listed above and to branch out on your own once you have tried this version. You may wish to

thicken the stew, if so now is the time to do so. You can either use cornflour suspended in cold water, or you can prepare a dark roux and use that. In either case you will find detailed instructions in the sauces chapter (Chapter 9).

Once you are happy with the flavour make some dumplings (see below for instructions) and add them to the stew and cook for another 20 minutes before serving. NB you need to make sure there is plenty of liquid in the stew as the dumplings will absorb a good deal so you should add enough boiling water to the stew so that there is a layer at least 1cm deep above the meat in the pan before adding the dumplings.

Dumplings

Mix the self raising flour and suet together with a little salt in a good sized mixing bowl (about 2 pints at least). add a little of the water (no more than 20 ml) and stir the ingredients together. Add a little more water and stir again. Keep on going until the mixture just starts to stick together – it should still be quite dry. Now, with your hands compress the mixture into a ball and gently knead it into a solid mass. Next divide it into 8 portions and roll them between your hands into smooth balls.

Key points in the recipe
- **Brown the meat** – remember flavour molecules are small molecules – heating the large proteins above ca. 140 °C starts breaking them down into smaller molecules that have flavour.
- **Deglaze the frying pan thoroughly** – the flavour comes mostly from the Maillard reactions and many of the reaction products remain in the pan.
- **Thicken the stew** – making a stock with vegetables provides starch to thicken the stew – the 'thick' texture makes the liquid coat the mouth and gives the stew 'richness'.
- **Taste regularly and often** – keep adjusting the flavour until it is just right.

Why the recipe works

The stock is made with several 'starchy' vegetables (carrots, parsnips and potatoes). During the long cooking much of this starch is released into the stock and even more is added when the vegetables are passed through the sieve at the end of cooking. The starch is in the form of 'granules' that can swell up to many times their original volume (potato starch granules can swell up to 100 times their original volume in hot water); these swollen granules will help to thicken the stew.

Cutting the meat into small pieces ensures that it will fall apart easily in the mouth and seem quite tender even though the muscle proteins have become coagulated and tough. The connective tissue between the muscle proteins breaks down into very soft material and allows the fibres to be easily pushed apart when chewed. So provided the bundles of muscle fibres are not very long the meat appears very tender. Cutting the meat into small pieces makes sure that the bundles of fibres cannot be longer than the pieces of meat are thick.

Problem	Cause	Solution
Not thick enough	Insufficient starch in stock (probably didn't sieve the vegetables well enough)	Next time make a 'thicker' stock in the first place – either use more potatoes and parsnips in the stock, or make sure you scrape more of the vegetables through the sieve at the end of cooking. This time thicken with a dark roux made with the fat left from sautéing the onions
	Very little gelatin in the meat	Next time use meat with more gristle, etc. This time thicken with a dark roux made with the fat left from sautéing the onions
Too thick	Too much water evaporated during cooking	Next time check more often and add more water, or use a pan with a better fitting lid For now remove the dumplings and add some boiling water and stir the stew until it reaches a suitable consistency.
	Too much liquid absorbed by the dumplings	Next time add more water before putting the dumplings in. Try making dumplings with less flour and more suet. For now remove the dumplings and add some boiling water and stir the stew until it reaches a suitable consistency.
Meat tough	Not cooked long enough	Keep on cooking!

Variations on the basic recipe

Boeuf Bourguignonne

Boeuf Bourguignonne is a traditional dish made from high quality meat with little connective tissue, that is nevertheless cooked for a comparatively long time. It needs to be cooked for long enough to break down the collagen holding the muscle bundles together, but not so long that any elastin or reticulin in any tendons or cartilage is denatured. This combination of a quality cut and a long cooking time gives a characteristic, "dry", texture to the meat, using a cut with lots of connective tissue gives, by comparison a "greasy" texture. You can, of course, make this recipe using any cut of beef, although if you use a cut with a great deal of connective tissue you will need to increase the cooking time.

Ingredients:

500 g	Sirloin
200 g	Bacon – preferably unsmoked back
16	shallots

Two cloves of garlic, crushed

Salt and pepper to taste

1.5 litres	stock (prepared as in beef stew recipe)
0.5 litres	red wine – preferably Burgundy or failing that any wine made from the Pinot Noir grape

Method

Prepare the stock as for the Beef Stew recipe above. Cut the bacon into 2 cm squares and the sirloin into cubes about 1.5 cm in size. Fry the sirloin in a little oil or dripping, until well browned a few pieces at a time – again following the instructions in the stew recipe. Put the browned meat in a large metal pan – preferably one that can be used on the hob as well as in the oven. Once all the sirloin has been cooked fry the bacon in the same way and add that to the sirloin in the large pan. Next, deglaze the frying pan with some of the wine. Then add the rest of the wine, the crushed garlic and 1 litre of the stock to the pan with the meat and bring to a boil. Let the mixture boil well for a minute or two to get rid of the alcohol from the wine and then cover and reduce to a simmer for 10 minutes. Check to see that all the meat is well covered with liquid and if necessary add some more stock, taste the liquid and add pepper and salt a little at a time to your taste. If your pan is oven proof then put it in the oven at 170 °C otherwise transfer all the meat and stock to a casserole and put that in the oven. Leave to cook for about one hour. Meanwhile, peel the shallots and gently sauté them in a little oil or butter until they just begin to brown. Check the level of the liquid in the casserole every 20 minutes and add more stock if needed (the liquid level should be sufficient to cover all the meat). Add the shallots about 20 minutes before you want to serve the dish (which should spend about one and a half to two hours in the oven altogether). It is traditional to serve Boeuf Bourguignonne with mashed potatoes and fresh green vegetables in season.

Oxtail soup

Ingredients

1 Oxtail (about 800 g)

2 – 3 litres stock

Salt and pepper to taste

A little dripping or oil to fry the oxtail

If possible get your butcher to cut the oxtail into segments and then to split the larger pieces through the middle of the bones. If this is not possible you should try to chop through the bones as best you can to cut the tail into pieces as small as practical. Once the oxtail has been cut up you can either fry it (as for the stew and Boeuf Bourguignonne recipes) or roast it to develop the Maillard reactions and bring out the flavour. I find it is often best to roast the tail for about an hour at 200 °C and then fry it later. Whichever method you choose, remember to deglaze the pans with some of the stock. Once the oxtail has been fried or roasted put it, together with the stock used to deglaze the roasting and frying pans and about 2 litres of additional stock into a large sauce pan and bring to a boil. Once the pan has come to the boil, turn down the heat and leave to simmer for at least 3 hours (preferably up to 6 hours). During this simmering the meat will come away from most of the bones in the oxtail and the connective tissue in the bones will dissolve and thicken the soup.

To finish the soup, remove the bones (scraping any remaining meat from them) and then simply season to taste, and thicken if necessary using flour, or cornflour as described in Chapter 9 (Sauces).

Some experiments to try at home

Here are a few experiments that you might like to try out at home to illustrate the main scientific principles discussed in this chapter. I always find people understand better once they have tried these sorts of experiments for themselves; it is far better to see for yourself than simply to believe everything you read!

1. An experiment to see how meat gets tough when cooked

Take a slice (about 1 cm thick) of good quality steak (sirloin, or rump, for example), trim it well and cut into about 10 roughly equal sized pieces. Now cook these pieces for different times – either in a frying pan, or under a grill. Make sure that all are treated in, as far as possible, exactly the same way.

Begin by putting one piece on to cook, then after 2 minutes add a second, and after another two minutes add a third, and so on until all the pieces are being cooked. Turn all the pieces regularly (say every minute). After the final piece has been cooking for 2 minutes take all the pieces off and test them for toughness and flavour. Make sure you know how long each piece was cooked for.

You can use both objective and subjective tests for the toughness. An objective test would be to see how deeply a blunt knife will cut into the pieces with a light pressure. The deeper the cut the more tender the meat. A subjective test is to try chewing each piece.

Record your results, along with your thoughts about the colour and flavour of the different pieces. Hopefully you will find that the pieces cooked for comparatively short times (up to say 8 minutes) are all quite tender, while those cooked for longer become increasingly tough. At the same time you should find an increasing depth of flavour with the cooking time. Once you have decided on your personal favourite you should find it a relatively simple matter to ensure you cook your own steaks to that standard in future.

The toughening occurs in two stages. The first stage is caused by the muscle proteins that are extended along the muscle fibres contracting into coils as they denature. The pieces of meat should be getting thinner during this stage as the fibres contract – can you see this happening? The second stage is the coagulation of the now denatured muscle proteins into lumpy, knotty masses. There should be a much greater degree of toughening in this stage, but no noticeable changes in the shape of the pieces of meat.

2. An experiment to see how collagen turns into gelatin

For this experiment you will need some meat with lots of connective tissue (e.g. oxtail, or shin of beef) preferably still connected to the bones with tendons, etc. Cut the meat and tendons into small pieces and cover with water, bring to a boil and leave to simmer making sure the water does not all evaporate. Every half hour remove about 50 ml of the water (and top up the pan with more boiling water). Put each lot of removed water in a separate glass and place in the fridge.

You should find that the water removed after about 2 hours simmering will thicken on cooling, and the one removed after about 3 hours should form jelly as it cools. Note, these times will vary a great deal depending on the actual amounts of connective tissue in the meat you are boiling.

As you heat the collagen in the meat and tendons above 60 °C it slowly starts to change from its triple stranded helical form into a single stranded form that is known as gelatin. The gelatin is soluble in water and so the concentration of gelatin slowly increases as more collagen denatures. When cold, gelatin molecules aggregate together in a loose structure that can bind a lot of water (see Chapter 2). At low concentrations of gelatin this leads to a 'thickening' effect; while at higher concentrations it leads to the formation of jellies.

3. An experiment to see how the browning reaction is affected by the temperature

For this experiment you will need several small pieces of steak; you will simply cook these at different temperatures until the centre is 'pink', and then see how much flavour has developed. To cook the pieces at well defined temperatures, use the oven. Put a baking sheet in the oven and let it heat up. Once it has reached the oven temperature, then put the piece of meat on the tray and cook for the time given in the table below before taking it out (remember to turn the meat once during the cooking). Next raise the oven temperature to the next highest temperature and repeat.

You should find that the meat has very little flavour when cooked at temperatures below 130 °C, and that the 'meaty' flavours develop and intensify as the temperature is raised to about 180 °C. As the cooking temperature is raised higher, the flavour changes and you should start to detect more 'burnt' than 'meaty' flavours.

The Maillard reactions have different end products depending on the temperature (and to some extent the time) they occur at. There is a fairly narrow temperature interval (140 – 180 °C) which seems to provide those flavours most of us prefer.

Oven Temperature	Cooking Time
100 °C	12 minutes (6 minutes each side)
120 °C	9.5 minutes (half of the time on each side)
140 °C	8 minutes (half of the time on each side)
160 °C	7 minutes (half of the time on each side)
180 °C	6 minutes (half of the time on each side)
200 °C	5.5 minutes (half of the time on each side)
220 °C	5 minutes (half of the time on each side)
240 °C	4 minutes (half of the time on each side)

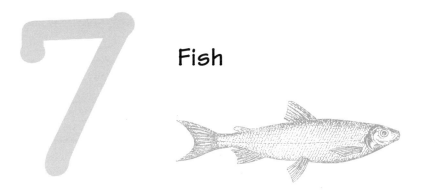

7 Fish

Introduction

The water in which they swim supports fish so, unlike land based animals, they do not have to support their own weight. This support from the water allows fish to have a different arrangement of muscle proteins from mammals. The muscles need to work much less hard and are not required to be able to exert such large forces. Consequently, fish muscles are generally much weaker than those of

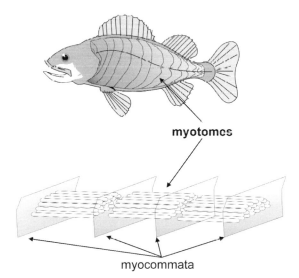

Figure 7.1. Fish muscles are different from those in mammals. The fibres (myotomes) are very short and are separated by delicate membranes (myocommata)

mammals. The proteins in fish muscles are not arranged in long fibres running right through the muscle, but rather they are organised into short bundles which are joined together by delicate membranes.

The main consequence of the difference between meat and fish muscle tissue is that there is no tough connective tissue between the muscles and the bones. So there is no need to cook fish for any length of time to make it tender. In fact, fish rarely needs to be cooked at all – we simply prefer it to have been cooked a little to soften the texture a little. However, if fish is cooked for any length of time, it will tend to fall apart as the tissue between the muscle fibres is very easily destroyed by heating.

The Smell of Fish

We are all familiar with the smell of fish, however, it is less well known that freshly caught fish have no odour at all. The characteristic "fishy" smell develops as chemical reactions take place in the flesh of the fish sometime after they have been caught. Fish are "cold-blooded", that is to say they do not control their body temperatures but rather operate at the temperature of their surroundings. Many fish live in cold seas, so that the enzymes they use to metabolise food need to be active at low temperatures (sometimes as low as 4 °C). The enzymes in meat, by contrast, work at body temperature (usually 36 to 38 °C) and are only very slightly reactive at low temperatures. The end products of some enzymatic reactions will start to accumulate once the fish is dead and its circulatory system is no longer functioning. It is these waste products that give fish its characteristic smell.

Further, off flavours and smells are introduced by the action of bacteria. Such bacterial reactions generally begin only after the end of rigor mortis which is usually after about 6 hours. However, if the fish is stored in ice, the end of rigor mortis can be delayed up to as long as a week, so keeping the fish fresh longer – this is particularly important for fish caught at sea and transported long distances before being sold to the consumer.

When you are buying fish it can be useful to remember that the fresher a fish, the less strong will be its smell. You should always eat fish as soon after it has been caught as is possible. Remember that the enzymatic reactions that can themselves consume the flesh cannot be delayed by refrigeration as they can for meats, so you need to eat the fish before its own enzymes eat it!

Cooking Fish

Most fish have subtle, delicate flavours, coming mainly from the oil in the flesh. Although it is possible to develop flavour through the Maillard reactions in the same way as is used when cooking meats, the resulting stronger flavours will mask the delicate flavour inherent in the fish, so in general little, if any browning of the outside of the fish is carried out.

In practice, cooking fish simply involves a little heating to help denature some of the tissue between the fibres and to heat the flesh to a more pleasant temperature for consumption. Various methods of heating are possible, and most are used in one recipe or another. The commonest techniques are to steam whole fish (using a special fish kettle), to sauté, or fry the fish in a little hot oil, or butter, or to bake the fish (usually in a sauce) in the oven, or to cook the fish under the grill. Deep fat frying is less common in domestic cookery, but still widely used in fish and chip shops.

Why is Fish White?
As we have seen in the previous chapter, there are two types of the protein myosin which can be found in muscle. These different types use different chemical routes to provide their energy.

Although both types occur in all muscle they can be considered as two distinct muscle types. In fact since the proportions can vary considerably it is not unreasonable only to consider the two types individually.

The 'slow' fibres burn fats to provide the energy; these muscles need oxygen to operate. The 'fast' fibres burn glycogen, and do not need to use any oxygen. Accordingly, muscles made from fast fibres do not need any myoglobin and are always white.

An important difference between fast and slow fibres is that the slow fibres are well suited to working continuously, while the fast fibres can only operate in short bursts.

All land animals need to support their own weight, hence they must have a lot of slow fibres in their muscles, which are usually dark. Fish, by contrast, have their weight supported by the water, thus they do not in general need any slow fibres in their muscles.

Some fish, notably sharks, are denser than the water in which they live; they literally have to keep on swimming all the time to stay up. These fish need some slow fibres and in consequence have darker meat than most other fish. Others, such as salmon have pigments that colour the flesh.

The texture of fish muscles is also rather different from that of land animals, in that the fibres are rather short. The short fibres have a tendency to fall apart when heated, making it rather easy to over cook fish.

Key points to bear in mind when cooking fish
- Keep cooking times as short as possible
- Use only the freshest available fish
- Fish have delicate flavours – do not use strong flavours in accompanying sauces

Why are these key points important

The muscle proteins in fish are very delicate and easily denatured, cooking for long times will make the texture very soft and the fish will begin to fall apart. The natural enzymes in fish operate at quite low temperatures, so the degradation of the fish by its own enzymes will begin immediately after death and continue even at quite low temperatures.

Methods of cooking Fish

There are as many ways of cooking fish as there are ways of cooking anything! Cooking fish is really rather easy, the only problem is that overcooking fish is probably even easier! The methods described below are all often used, and are listed in the order of increasing likelihood of overcooking being a problem.

Steaming and Poaching

Steamed fish is entirely cooked in an atmosphere of steam (normally using a special fish kettle). Water is boiled to produce the steam at the bottom of the steamer, while the fish sits on a perforated platform above the water level so that the steam always surrounds it. Poaching is carried out with the fish partially in water and partially in steam. Usually, the fish is placed on a bed of vegetables, in a casserole or metal pan, and water or stock is added to cover the vegetables. The pan (with its tightly fitting lid on) is then either heated in an oven, or on the top of the stove.

All types of fish and seafood can be steamed or poached to produce good results. The main advantages of steaming are that the temperature never rises above 100 °C, so that there is no risk of burning; and that the fish is kept moist by the atmosphere of water surrounding so there is little risk of serving a dried out dish.

Since the temperature of cooking is no greater than 100 °C, the Maillard reactions, so important for flavour development in meat cookery, will never occur and the flavour is simply that of the fish itself – mainly coming from the oils in the fish.

Problems that may occur when steaming or poaching fish

Problem	Cause	Solution
The flesh falls apart during cooking. The fish falls apart when I take it out of the pan.	You have cooked the fish for too long a time.	Next time cook for a shorter time – you can test the fish by gently pressing it with a spoon. As soon as the spoon dents the fish it is ready – cooking any longer will make it break up.
The fish is tough.	You haven't cooked the fish long enough.	Next time cook a little longer – see above for testing when the fish is cooked.

Shallow Frying or Sautéing

Sautéing in butter or oil is probably the commonest method of cooking fish used in the home. The outside of the fish can get hot enough to allow the Maillard reactions to occur and develop a little new flavour. Care needs to be taken to ensure the Maillard reactions do not proceed too far, as the fish will quickly develop a burnt flavour, which while it may sometimes be acceptable in meat cookery is overpowering in fish.

Both fillets and small whole fish can be sautéed successfully, but when cooking fillets the risk of burning is greater so it is common practice to coat the fillets with some flour, or ground nuts, etc.

In general, it is preferable to sauté fish in good oil, rather than in butter, as butterfat tends to oxidise (burn) at a lower temperature than saturated oils such as Olive oil. Thus, the risk of developing unpleasant, burnt, flavours is greater when using butter to fry fish.

Problems that may occur when sautéing fish

Problem	Cause	Solution
The fish falls apart during (or after) cooking.	You have cooked the fish too long.	Next time cook for a shorter time. Judge when the fish is cooked by pressing gently – the fish should recover quickly, if it remains dented it will soon become overcooked.
The fish has a burnt taste.	You have allowed the Maillard reactions to proceed too far. Either by cooking too long or by cooking at too high a temperature.	Next time cook for a shorter time, or at a lower temperature. If the flesh of the fish is moist and does not fall apart, then you probably used too high a temperature – if the flesh is at all dry and falls apart, then you probably cooked it for too long.
The fish is tough, rather than tender.	You have not cooked the fish long enough.	Next time cook the fish a little longer. If the outside of the fish is well browned, then use a lower temperature next time. For now cook the fish a little longer.
The fish has a rancid flavour	Either the oil you used to cook the fish was old and oxidised, or the fish was not fresh.	Make sure you use fresh fish and oil.

Deep Frying

Deep-frying is the most traditional, British, way of cooking fish – it is the basis of our "Fish and Chips". The basic idea is to coat the fish with a thick, almost solid, batter which will cook very quickly in the hot oil and seal in the moisture in the fish. In practice, the fish then cooks in steam from it's own juices so that the temperature remains below 100 °C, thus producing a tender and moist product.

However, the risk is that there may be a small break in the batter which can allow some hot oil to penetrate into the fish and allow the steam from the fish's own juices to escape in which the fish can become seriously overcooked and completely dried out.

When deep-frying fish, the batter is of paramount importance. The batter needs to be quite thick before coating the fish (usually filleted). To achieve a thick batter use roughly equal weights of liquid and flour. To ensure the batter sets easily, use some beaten egg in the batter. The remaining liquid can be milk, water or even beer, depending on taste. The batter should be left to stand for an hour or so to allow it to thicken as the starch in the flour absorbs the liquid. The fish fillets should then be completely immersed in the batter and left on a dry, lightly floured surface for a few minutes, to allow the batter to set as some of the water evaporates, before being put in the hot oil and fried.

Problems that may occur when deep frying fish

Problem	Cause	Solution
Batter is not crisp.	The temperature of the oil is too low, or the cooking time is too short.	Make sure the oil is at the right temperature (180 °C), and make sure you do not put too many pieces of fish in the hot oil at one time (when you put the fish in the oil the temperature drops – putting several pieces in at once makes the temperature drop considerably).
Batter is not "fluffy"	The batter was too stiff, or the battered fish was left too long before frying.	Prepare the batter with a little more liquid next time.
The flesh of the fish is dry.	The fish was cooked for too long, or it was not fully encased in batter, or the batter fell off the fish.	Make sure the fish is fully covered with batter before it is put in the hot oil. Reduce the cooking time.
The flesh has fallen apart.	The fish was cooked for too long, or it was not fully encased in batter, or the batter fell off the fish.	Make sure the fish is fully covered with batter before it is put in the hot oil. Reduce the cooking time.
The fish tastes rancid	The oil has gone rancid from repeated use, or the fish was not fresh.	Make sure you replace your cooking oil regularly and often, make sure you use fresh fish.
The fish falls apart in the hot oil.	The batter has not properly stuck to the fish, or the fish was not fully coated, and the fish was cooked too long.	If the batter falls off the fish then make a thicker batter and make sure you dry the fish thoroughly before coating it. Reduce the cooking time.

Baking and Roasting

Cooking fish in the oven at high temperatures can lead to significant flavour development through the Maillard reactions, but can also lead to loss of moisture and to very dry tasting fish, as well as carrying the risk of the fish becoming overcooked and falling to pieces.

In his book, "Fruits of the Sea", Rick Stein makes a useful distinction between baking and roasting fish. Baking involves cooking the fish in a deep pan at a moderate oven temperature (180°C), often along with some vegetables, so that some of the steam that rises from the fish and vegetables remains above and around the fish as it cooks. Keeping the fish in an atmosphere of steam helps to reduce the overall loss of water from the fish and retaining a moist product. Roasting is the process of putting a large piece of fish, or a whole fish on a trivet over a shallow pan and exposing it to the fierce heat of a hot (230°C) oven. When cooked properly, the fish skin will crisp up well providing a good combination of textures from the crisp outer layer to the moist and tender flesh inside the fish. However, the risk of overcooking is very great so this is a method that needs plenty of practice to get the timing just right. The variability of domestic ovens is so large that it is not possible to provide any accurate guide to cooking times for roast fish. You simply have to test to see when the skin becomes crisp and to try to tell, by gently pressing on the fish when it starts to become dry. The idea is to let the skin crisp up, but to prevent the flesh becoming at all dry.

Problems that may occur when baking or roasting fish

Problem	Cause	Solution
The fish is rather dry and/or falls apart.	The fish has been cooked too long or at too high a temperature.	Next time cook for a shorter time at a lower temperature.
The fish has a burnt taste	The outside has 'browned' too much.	Either cook for a shorter time, or protect the fish from browning by covering with foil for part of the cooking time.

Grilling

Grilling a piece of fish is probably the quickest and simplest way to cook it. In a well grilled fish the high temperature at the surface activates the Maillard reactions generating flavour, while the very speed of cooking prevents the middle becoming hot enough to damage the delicate texture of the fish.

However, grilling is also the method most fraught with risk of ruining a perfectly good piece of fish. Ideally the grill will be very powerful, and the fish is cooked for just a couple of minutes. If the grill is less powerful, or if the fish is left under the grill for more than a minute or two, then the denaturing of the fish muscle proteins will lead to their coiling up and the expulsion of water so giving a very dry and even tough product. As the cooking time increases, so the depth into the fish where the muscle proteins have denatured increases and ever more moisture is removed. Many domestic grills are not very powerful, so by the time that the outside has browned and the flavour developed, the inside of the fish may well be thoroughly overcooked.

For successful grilling of fish you really need a powerful grill – most gas cookers do have good grills, while those on electric cookers tend to be inadequate for this purpose. A solution, but one that needs a steady hand and some confidence, is to use the extreme heat from a blowtorch to grill your fish. Put the fish on a baking sheet, light your blow torch and then bring the flame to the fish until the fish just browns – keep the blow torch moving over the fish surface all the time. Take great care not to linger in one place, or the fish will burn. As soon as the fish is lightly browned on one side, put the blowtorch down, turn the fish over and repeat on the other side. The whole process should take no more than a minute or two.

Problems with grilling fish

Problem	Cause	Solution
The outside is not crisp.	The grill is not powerful enough or the cooking time is too short.	If the flesh of the fish is not dry, or falling apart, then cook for longer next time. Otherwise, if your grill is not powerful enough for this cooking method to work, then you can use a thicker piece of fish next time, or simply use a different cooking method!
The fish falls apart The fish is dry	The fish is overcooked.	Try a shorter cooking time or use a thicker piece of fish next time. If this means the outside does not become crisp then your grill is not powerful enough for this method of cooking.
The fish has a burnt taste	The fish was cooked too long, or the grill was too powerful.	Cook for a shorter time, or with a lower setting on the grill.

Fish and Chips

Ingredients

1	Cod fillets (about 120g each)
100 g	Plain Flour
80 ml	Beer (preferably Guinness)
5 ml	Salt
1	Egg (large)
150 g	Potatoes peeled and cut into pieces about 5cm long by 1cm by 1 cm

Method

Beat the egg and add about half the beer to it. Put the flour and salt in a bowl and add about half the beer and egg mixture. Beat until it becomes a smooth paste adding a little more of the liquid if necessary. Slowly add the remaining beer and egg mixture beating all the time. Add more beer until the batter has the consistency of thick cream. Leave the batter to stand for about an hour. Dry the cod fillets on kitchen paper and put them one at a time into the bowl of batter. Make sure the cod fillet is well covered with batter. Take the fillet out of the batter and put on a floured board, turn over so that both sides are covered in dry flour. Repeat for each fillet.

Leave the fillets to stand for another 5 minutes and then put them one or two at a time into a deep frying pan of oil at 180°C to cook for 5 minutes each – put in a warm oven to keep warm until served. If you have two deep frying pans cook the chips at the same time as the cod, otherwise cook the chips before the cod in the same oil – put the uncooked potatoes in the oil at about 170°C for about 6 minutes and then increase the temperature to 190°C for a further 6 minutes until the chips are golden brown. Remove from the oil and dry on kitchen paper before serving. Keep warm in the oven if necessary.

Trout with almonds

Ingredients

4	Rainbow Trout
150 g	Flaked Almonds
100 g	Butter
Juice of one lemon	

Method

Ask your fishmonger to clean the trout and remove the heads if you do not want to serve them with their heads on (many people do not like a trout's eye staring up at them from their dinner plate – but it doesn't bother me!). Wash the trout in clean water and dry on paper towels. Melt the butter in a frying pan and heat gently. Lay the trout in the pan and cook gently for about 5 minutes on each side. The skin should just begin to turn a golden colour as you cook the trout.

Pour off any surplus butter and pour the lemon juice in the frying pan and let it boil away for a few seconds before taking off the heat and serving.

If possible cook all the trout at the same time using two or more pans. If you cannot do this then you will need to keep the trout cooked first warm while cooking the rest. However, remember that these trout being kept warm will continue to cook, so cook them for a slighter shorter time in the first instance.

Sea Bass with a Raspberry sauce

I know that at the beginning of this chapter, I suggested avoiding strong flavours when cooking fish, but here is a recipe that uses a small amount of raspberries to provide a sharp and flavoursome contrast to a fairly robust fish, Sea Bass. As with all rules, you should try to break them once in a while!

Ingredients

400 g	Sea Bass
200 g	Raspberries
5 g	Sugar
1	Lemon
200 g	Potatoes

Oil and butter for frying
$^1/_2$ teaspoon cornflour
A little cold water

Method

Prepare rosti potatoes by shredding the potatoes into thin strips about a millimetre thick and wide. You can do this with a sharp knife, or with a lemon zester, or using a potato peeler to make wide strips and then cutting these into thin strips with a knife. Put a little oil in a small frying pan (ideally about 10 cm in diameter) and heat until almost smoking. Put in a layer of potato strips and press down well – they should stick to each other. After about 30 seconds turn the rosti over and brown on the other side. The rosti potatoes will be made from the lattice of the potatoes glued together by some of the starch that leaches out when the strips are cut; they should be thin (no more than 2 or 3 millimetres thick and well browned on both sides. Make up as many rosti potatoes as you will be serving pieces of Sea Bass.

Next prepare the raspberry sauce. Wash the lemon and cut the peel thinly, then extract the juice. Puree the raspberries and pass through a sieve to remove the pips. Heat the puree in a saucepan with the sugar and about a quarter of the lemon juice, allow to boil for a few minutes to reduce to about 2/3 of its original volume. Thicken with a half teaspoon of cornflour which has been beaten into about 20 ml of cold water. Take off the heat. Sweeten to taste, if you wish.

Finally cook the Sea Bass. Cut the Sea Bass into 4 steak sized pieces and quickly brown on both sides in smoking hot butter in a frying pan. Put the Sea Bass in a dish with a tightly fitting lid with the remaining lemon juice and simmer for 5 minutes. To serve, put each piece of cooked Sea Bass on a rosti, in the centre of a large plate and spoon over a little of the raspberry sauce. Decorate the rims of the plates with the remaining raspberry sauce and the lemon peel. Serve with fresh green vegetables of your choice.

Crab soup

This recipe makes a very rich, thick crab soup suitable for serving as a starter. Since the recipe calls for a lot of crab meat – to provide the flavour, it can be a rather expensive dish.

Ingredients – for 4
1 large crab, dressed, or 200 g crab meat
1 large carrot
1 medium onion
300 ml fish stock
100 ml dry white wine
50 g plain flour
160 g butter
50 ml double cream
20 ml tomato puree (about 1 tablespoon)
About 1 teaspoon chopped fresh coriander (or parsley)

Method

Separate the coral from the rest of the crab meat (if you are using crab meat and have no coral set aside about 75 g of the meat). If you are not familiar with shellfish the "coral" is the grey – green stuff that looks like a lot of small soft balls. Chop the carrot and onion into small pieces (about 5 mm in size). Put the carrot and onion in a heavy bottomed pan on a very low heat with no liquid added, cover, and allow to sweat for about 5 minutes. Add the crab coral, or the meat you set aside, the tomato puree and the fish stock. Leave, covered, to simmer for an hour, check from time to time that the vegetables are covered with liquid and add a little water if necessary. Meanwhile, sauté the remaining crab meat in half the butter.

Pour off the stock into another pan and pass the solids through a sieve – you should aim to reduce the amount of solids to much less than half during the sieving. Take care to pass as much as possible through the sieve as much of the flavour and texture of the soup comes from these pureed solids. Use a spoon to work the solids around the sieve and to make them into a puree fine enough to pass through. All the carrot and the crab coral should go through the sieve – you should only be left with some of the onion. Measure the amount of liquid you have – if it is more than 500ml then reduce it, by boiling to about 450 ml.

Prepare a soft roux with the remaining butter and the flour in the pan you used to sauté the crabmeat. Simply melt the butter and then add the flour and stir on a low heat for about 2 minutes – allowing the flour to attain a light golden colour. Take the pan off the heat and stir in a little at a time the liquid that was passed through the sieve. Make sure you keep the texture of the mixture very smooth – do not allow any lumps to develop. Now transfer the mixture to the saucepan, add the cooked crab meat and the white wine, bring to a boil and cook for a further 3 minutes. You should now have a very thick, pinkish coloured, soup. Finish the soup by stirring in the double cream and immediately serve in small coffee cups with a sprinkle of chopped coriander on top.

Preserved Fish

Fish, as we have already seen, is particularly susceptible to decay. The enzymes in fishes' bodies continue to remain active at quite low temperatures, so that even in a refrigerator, fish will deteriorate quickly. Of course, with modern cold transport methods and careful stock control, fishmongers are able to provide us with fresh fish even large distances from the sea where they are caught.

However, this has not always been the case. Many years ago, there were no mechanisms for keeping fish cold enough to prevent spoilage. For example, in the 1600's ships from my home town of Bristol traded in cod. However, when the cod was caught at sea either around Iceland or off the coasts of Newfoundland, the seafarers had no means to keep them cold enough for long enough time to bring them back to Bristol without them rotting on the journey.

Similar problems had faced people for many years and several solutions had been found. A variety of methods have been used traditionally for centuries all over the world, to preserve fish. At the heart of all these preservation techniques, is the fact that if the water content of the fish is low enough then enzymes cease to act, even in warm environments. Further, the low moisture content prevents any micro-organisms from attacking the flesh of the fish as they too need plenty of water to survive.

In practice, if the moisture content is maintained below about 13%, any food can be preserved more or less indefinitely.

The commonest, and almost certainly the original method of preserving fish, is to dry it. There are many different types of dried fish prepared around the world today. Many of these are regarded as delicacies. There are many different ways of drying fish, from the simple leaving out in the sun to the more complex filleting and drying on racks over an open fire processes used by the Bristolians in the cod trade.

A second common preservation technique, and one that was probably originally discovered when handling fish, is salting. Salt will draw moisture out of the fish, and also out of any micro-organisms that might attack it. So it is reasonable to suppose that at some time in the distant past a fisherman kept some of his fish caught at sea in sea water, and found that when the water evaporated, the layer of salt prevented the fish from rotting. Today, salt cod is a delicacy in parts of Italy.

The problem with dried, or salted fish, is that to be able to eat it most people need to increase the moisture content to make the fish chewable and get rid of any excess salt to make it palatable. Of course, there are those who think eating dried or salted fish as they are is quite acceptable and even special. I cannot agree.

The processes of reconstituting dried or salted fish are numerous and have much associated mystery history and ritual. The results are generally either rather mushy, or rather leathery, and often too salty for many palettes. Probably the most extreme example is the Norwegian Lutefisk (see the panel for a detailed description of this dish – my least favourite of all!).

One process of preserving can produce a fish that is not only edible with no need to treat it in any way, but is also generally regarded to enhance the flavours in a most positive manner. Smoked fish of all kinds remain moist, but nevertheless are protected by toxins in the smoke from enzyme action and from bacterial decay. Smoking, as a means of preservation was probably discovered at some time when fish was being dried over an open fire. The flavour of the smoked fish was probably enjoyed, so further experiments would have been tried out to improve the result.

My worst nightmare – Lutefisk!
It was the week before Christmas and my host determined that we should sample the traditional Norwegian Christmas fare at its best. So he booked a table at what he told me was the best Norwegian restaurant in Oslo for the two of us. It soon became clear that this was a man possessed of a cherished memory that he needed to relive. His wife, he told me, could not be bothered to take the great time necessary to prepare Lutefisk so he had not eaten any for many years now. He was extraordinarily happy that he now had a good excuse to visit a restaurant and eat Lutefisk – when I wondered why he didn't just go every year his face clouded over and he muttered something about the expense. I later gathered that the meal we ate that evening probably cost far more than any other meal I have ever eaten anywhere in the world!

As we talked about Christmas traditions around the world, about science and politics, I eventually brought the conversation back to the meal we were about to sample. I had never previously heard of Lutefisk so I naively asked what kind of fish it was, expecting a simple answer such as 'Oh its a fish that only lives in a few deep fjords in the north of Norway' or some other justification for the obvious expense and rarity of the dish. Instead, much to my surprise, my host simply told me it was a form of dried cod. Dried cod, I thought, why so rare, why so special? So I asked "how is it prepared?" Just at that moment the food started to arrive at the table and I was soon to find out more than anyone could possibly want to know about Lutefisk.

As the waitress piled dishes on the table I started to wonder what I had let myself in for. With the appearance of a large dish containing a dirty green mush with the texture of badly mashed potatoes I started to become a little concerned. When this was followed by an even larger dish of what really was mashed potatoes my worries grew. However, it was the enormous bowl of hot bacon fat with a few pieces of chopped fried bacon floating around that finally made me realise that, even if this was going to be the best meal of my host's life, the chances that it would be the best I had ever eaten were extremely small.

Finally, with some ceremony, the waitress brought out the two oversized plates piled high with the Lutefisk. At first I was baffled, where was the cod? My plate was covered with a mound of a translucent white gelatinous substance. I presumed this was a sauce on top of the fish. But no, how wrong could I be? The wobbling mound was the cod. What could they possibly have done to a perfectly good piece of cod to reduce it to this sorry state I wondered?

The process is simple, my host told me; first they catch the cod in the summer then put them on a rack in the summer sun to dry out. The dried fish are then stored as piles of bricks until needed (often years later). The older the better, as they say. Then at Christmas time a brick of this dried Stockfisk is taken from the store, and rehydrated with the aid of the lye or caustic soda before being cooked and served up as the wonderful Christmas dinner.

Many years ago in the North of Norway the main diet was fish, mostly cod. However, although cod were plentiful in the short summer months they were hard to find in the winter when the fjords froze over and fishing became difficult. To solve the winter food shortage the people needed to find ways to preserve fish caught in the summer months to eat in the winter. The first and most obvious solution is simply to dry the fish in the sun. But then you need to rehydrate the fish before eating it. The Vikings must have discovered that if they used water to which they had added ashes from their wood fires this rehydration was quite fast.

Ashes from wood contain large amounts of sodium and potassium carbonates (that's why these mildly alkaline chemicals are known as soda-ash and potash). Alkaline solutions (or lye) have many uses – the commonest today is as drain cleaner. Strong alkalis, such as sodium hydroxide, are powerful chemical reagents. Alkalis react with fats and oils to produce the fatty acids which are more commonly known as soap. Indeed the Romans used a mixture of sodium and potassium hydroxides and carbonates to make soap from animal fats. Alkalis also react with proteins, at first the proteins are denatured, that is they lose their shape. But after long exposure, the proteins can be broken down into smaller molecules.

When a dried cod is soaked in an alkaline solution (these days usually sodium hydroxide in water) it will undergo lots of changes as it is rehydrated. The oil in the skin and between the muscles is quickly converted into soapy fatty acids, giving the finished product a soapy taste (even after most of the fatty acids have been washed away in the subsequent soaking process). The muscle proteins are denatured and lose their firmness, in a process similar to cooking the fish – long exposure to the alkaline solution is very similar to prolonged boiling of the fish. The connective tissue, the delicate membranes between the muscles, are destroyed, which would under different circumstances lead to the disintegration of the whole fish. However, some of the molecules in the connective tissue as well as some of the denatured proteins undergo further chemical reactions and form a very soft 'jelly' (in a manner analogous to softly cooked egg yolks). It is this 'jelly' that holds the Lutefisk together while it is soaking in the alkaline solution. Even the bones will be softened and turn to a gelatinous consistency on prolonged exposure. As all these separate chemical reactions proceed so the structure of the fish is changed and it begins to form a homogeneous glutinous mass.

Of course, how far these chemical processes proceed depends on the strength of the alkaline solution, its temperature and the length of time the fish is kept soaking. It is the control over these variables that gives the producer of Lutefisk control over the final texture and taste of his product. As soon as the producer believes the fish are sufficiently rotted in the alkaline solution he removes them and rinses the fish in fresh water to remove the lye.

Much of this chemistry went through my mind as I looked at the plate of Lutefisk before me and wondered just how I was going to get through it without being rude to my host, a powerful man at the head of a large chemical company from whom I was hoping to extract some money to fund my ongoing research.

My host soon led the way quickly tucking in to large portions of the Lutefisk mixed with helpings of the potatoes, the green stuff (which turned out to be nothing more exotic than mushy peas) and the bacon fat, and then washing the lot down with a swig of aquavit.

I soon found that I could just about tolerate the slightly soapy flavour of the Lutefisk – it was the peculiar gelatinous texture that I found hard to cope with. The equally soft textures of the partly mashed potatoes and the peas did not help, but surprisingly, the somewhat salty bacon fat gave just enough edge to the mixture to make it possible to swallow without wincing. In practice it was only the swig of aquavit after each brave mouthful that kept me going – albeit rather slowly. My host had soon polished off all his plateful while I still bravely fought on. I tried to give the impression I always ate this slowly to make sure I savoured each moment of the experience. I believe I almost convinced my host that I was actually enjoying the meal – I did not really lie to him, I merely said I had never before eaten anything like it!

Eventually, after a mammoth effort, I managed to finish the plateful of Lutefisk and most of the side dishes. As soon as I had taken the final mouthful our waitress appeared at my side and whisked away my plate together with that of my host. My relief at seeing the end of the Lutefisk was, however, very short-lived. No sooner had she gathered up the empty plates than she pronounced in a lilting Scandinavian accent the words that I shall never forget as long as I live. "And now would you like your second plateful of Lutefisk?"

My host immediately accepted with such firmness that it took all the courage I could muster to decline the offer. The poor waitress was quite crestfallen. But I explained that I had already had a large lunch and simply could not eat anything further. My host's second plate arrived almost immediately and I sat in wonderment as he rapidly demolished his second helping with enormous relish.

By this time, having consumed I know not how much aquavit (did I mention that every time the bottle was emptied another magically appeared?), I needed a coffee before venturing out into the cold to find my way back to the hotel. Our waitress seemed surprised that I could drink coffee as I was so full. Nevertheless she did agree to bring a pot of the strongest coffee I have ever tasted. As we drank our coffee my host decided to round off his wonderful meal with a massive and most smelly cigar. I found the combination of heady cigar smoke, strong coffee with the fresh memory of the Lutefisk almost impossible to bear, but, from somewhere, I found the strength to keep my food down and listen to my dear host telling me how that was the best Lutefisk he had ever eaten and how lucky I was to have had such a good introduction to the Norwegian national dish!

Somehow, I'll never know how, I managed to escape from the restaurant without offending my host, and made my way back to my hotel in the freezing blizzard outside. The next morning, we were supposed to be having a 'wrap-up' session before departing for home. All I remember is one of my Norwegian colleagues banging on my door shouting that the meeting had started and where was I?

Smoked salmon soufflé

Ingredients (for 4 servings)
4 large eggs
120 g smoked salmon trimmings
80 ml dry white wine
25 g plain flour

Method

For a detailed discussion on preparing soufflés see Chapter 12. Begin by separating the eggs and chopping the smoked salmon trimmings into small pieces.

Next make a smooth paste of the smoked salmon together with the wine and egg yolks, using an electric blender, or by passing through a fine sieve. Thoroughly grease four 7 cm diameter soufflé dishes and preheat the oven to 180 °C. Beat the egg whites until they are very stiff and then fold the egg foam into the smoked salmon paste a little at a time. Finally spoon the mixture into the soufflé dishes and cook in the oven for about 10 minutes. The soufflés should rise about 3 cm above the tops of the dishes. Serve immediately, before there is any risk of the soufflés collapsing.

An experiment to illustrate the buoyancy of fish

In this experiment you will make a "Cartesian Diver". The idea is to make a simple device that will just float in water, but which under a little external pressure will dive. The principle is very simple. One part of the "diver" contains some air. If we put this under pressure, so the volume decreases but the overall weight stays the same, so the density increases.

If a body has a density less than that of water it will float, if the density is greater than water it will sink. If the density is exactly the same as water it will float under the water. Fish have a "swim bladder" filled with air, by contracting muscles and changing the density of that air they can control their buoyancy and determine how deep in the water they float.

To make your own "Cartesian Diver" all you need is a small piece of expanded polystyrene packaging material, some plastic material to make the shape of the diver and some pieces of metal (e.g. a few nails) and some plasticine to help adjust the overall density. Begin by cutting out of the plastic material the overall shape of your diver (say a penguin shape) and fold it over to make a hollow shape. Ideally the diver should be small enough to fit inside a 2 litre pop bottle. Then fix a small piece of expanded polystyrene inside the diver with some waterproof glue. Now float the diver on some water and see how much weight you need to add before it will just start to sink.

Fix some nails inside the diver with the plasticine until it only just floats. Now if you put the diver inside a pop bottle filled with water (and with the top securely screwed on) it will float to the top. But if you squeeze the bottle to apply pressure to the air trapped inside the polystyrene in the diver it should start to sink to the bottom of the bottle. You should be able to keep the diver at any height inside the bottle simply by controlling how hard you squeeze the bottle.

If the diver won't sink no matter how hard you squeeze the bottle take it out and add a little more weight, until it does work.

8 Breads

Introduction

The smell of freshly baked bread always makes a home feel welcoming. Bread fresh from the oven tastes so much better than anything from the bakery. So why is that so few of us cook our own bread these days? Bread is often perceived as being difficult to cook. This is a false perception; bread making requires no special skills. All that is needed to make excellent bread are the right ingredients and a little practice. It is true that the bread dough takes some time to rise before it can be baked so the process can be a little time consuming; however, the effort is well worthwhile.

In this chapter, (and Chapters 10 and 11) we will be concerned with various aspects of baking; breads (here), cakes (Chapter10) and pastry (Chapter11). At the heart of all these different baked good lies the flour and how it is treated. While a detailed discussion of the molecules (starches and proteins) in flour may be found in Chapter 2, together with some details of the internal structure of flour, we will recap the main, essential points in each of the 'baking chapters' as necessary.

Structure of Flour

Flour is made up from small starch granules. Each starch granule is in turn composed of a mixture of 'starch' molecules. In Chapter 2 you can find a more detailed description of these molecules, their structure and properties It may be helpful here to remember that there are two "starch" molecules – amylose (a 'linear' molecule) and amylopectin (a heavily branched molecule). Both these molecules are polysaccharides (long string molecules made up from lots of

Types of Flour

There are many different types of flour that you can buy. The names vary widely around the world and even within one country, making it difficult to give detailed definitions that are truly useful.

The first difference between flours is the type of grain they are milled from. All flours start life as cereal grains, these are then ground down (or milled) to a fine powder. In some flours the chaff, or outer skin of the grains is removed while in others (brown, or wholegrain, etc.) it is not. Grains commonly used for flour are wheat, rice, corn (or maize), barley, and several types of bean including soya, chickpea and fava. Although any of these different flours can be used in most recipes, with some adjustments, by far the commonest of these is wheat flour, and we will concentrate on wheat flour in the following.

There are many strains of wheat, and even a single strain will produce different proportions of starch and protein in the grains depending on the climate and soil conditions where it is grown. Flour manufacturers, like wine makers, often blend their flours to provide a more consistent product. There are two very different types of flour available: "Plain", or "All Purpose" Flours and "Self Raising" Flours. Self raising flours have raising agents such as baking powder added to them to assist in cake making. Details of how baking powder works are given later in Chapter 10. Plain flours have no additives (although some manufacturers add some "drying agents" to prevent the flour forming into lumps).

There are still several more different types of plain wheat flours. These are characterised by the amount of protein in the flour. Flours with high protein contents (above about 12%) are particularly useful for bread making and are often called "Bread" flours or "Hard" flours. Flours with a low protein content (around 8 to 10%) are termed "Cake" or "Soft" flours.

For general baking you need a moderate protein content of around 7 to 10% by weight. It is generally better to have a flour with a protein content at the lower end of this range for pastry (where the generation of gluten needs to be minimised). For breads, where the generation of gluten is essential, you need a higher protein content, over 10%, so special flours are essential. You should also be aware that some flours are sold as wholemeal, brown, or whole wheat, etc. These flours contain some of the chaff – or outer skins of the wheat grains and have a brown colour and distinctive flavour when used in baking. Most cooks prefer to use these whole flours to make breads – indeed many people prefer fresh bread to be 'brown' rather than 'white'. However, you must be very careful when looking at the protein content of these flours, since the chaff itself adds to the overall protein content without helping to provide any extra gluten formation. So you will need a higher protein content (say 13% or above) to be able to make a good bread with whole wheat flours.

sugar molecules all strung together). It is also important to remember that starch granules also contain some proteins the actual amount of protein depending on the source of the starch granules. The arrangement of the different molecules and their proportions determine the 'type' of flour (hard or soft, bread or cake, etc. – see panel) and how the flour will behave when used in baking.

For bread making the most important aspect of the flour is the formation of 'gluten' sheets as the wet dough is kneaded. When water is added to the flour the proteins on the outside of the starch granules rapidly absorb the moisture and become very 'sticky'. This process of absorbing water is technically referred to as "hydration". These sticky, or 'hydrated', protein molecules then begin to stick together, and so bind the granules to one another. If these bound granules are then moved apart the proteins between them become stretched. When they are stretched the proteins change their shape and interact with each other in different ways. The new interactions between the starch granule proteins lead to the formation of gluten. Gluten is not a protein by itself, nor does it occur in nature, rather it is formed when two different protein molecules (gliadin and glutenin) are made to interact with each other by the kneading of a wet dough, to form a "super protein" or "protein complex".

Gluten is a highly elastic material – the gluten as it develops forms thin sheets that behave rather like rubber balloons. In bread, these 'balloons' formed from the gluten sheets become 'blown up' by carbon dioxide gas generated by the yeast as the bread leavens, and thus make the bread rise. To make good bread you need these gluten sheets to be robust enough not to break as the carbon dioxide is formed and plentiful enough to capture the gas in very small bubbles. Large bubbles would lead to holes in the final loaf.

What is yeast

A text book definition might read: 'yeasts are single celled fungal micro-organisms that metabolise sugars into carbon dioxide and alcohol'. However, we do not need to have such a detailed, scientific picture. All you really need to know is that yeasts are very small living organisms whose food is mostly sugar and whose waste products are mostly Carbon Dioxide (CO_2) and alcohol. Over 160 different species of yeast have been identified. Different strains of yeast are used for different purposes. In baking, yeasts that produce little alcohol are favoured, while those that are able to produce large amounts of alcohol and can survive in alcohol rich environments are favoured in the wine making and brewing industries.

The conditions under which the fermentation occurs have a large effect on both the rate of fermentation and on the fermentation products. For example, the species of yeast used in almost all baking is *Saccharomyces cerevisiae*; this yeast will convert sugar (mostly glucose – see Chapter 2 for a discussion of the different types of sugar and where they can be found) into either alcohol and carbon dioxide, or in an oxygen rich environment into carbon dioxide and water. So when using this yeast to make beer or wine it is important to keep oxygen away, but when making bread it can be helpful if there is some oxygen around to help the yeast make more carbon dioxide.

The types of sugars present in the fermenting medium (in the case of bread – the dough) have a major effect on the progress of the fermentation. There are many different types of sugars, but the one that is favoured greatly by the commonly used yeast strains is glucose. However, if you add sugar to the dough you are actually adding sucrose, which has a different structure, so that the yeast must first convert it to glucose before it can metabolise the sugar into carbon dioxide. The yeast produces an enzyme, sucrase, that converts sucrose into glucose and fructose. Further, the yeasts can produce another enzyme, amylase, that can break down the starch molecules into maltose, another sugar that can be fermented into carbon dioxide and alcohol.

The temperature of the fermentation is also of major importance. While *Saccharomyces cerevisiae* will convert glucose into carbon dioxide even at temperatures as low as 5 °C, the rate of production of the gas increases exponentially as the temperature is raised up to around 38 °C. At temperatures over about 40 °C the yeast is slowly killed (it takes about an hour to kill off most of the yeast cells at 43 °C) and so less carbon dioxide is produced. There is no real optimum temperature for the fermentation, fermenting at temperatures below 20 °C is rather slow and the yeast needs a plentiful supply of glucose. However, the action of the sucrase enzyme that converts the sucrose into glucose is very slow at this temperature, so the fermentation is particularly slow. At temperatures above about 30 °C, other micro organisms can flourish and this can lead to yeasty, or off, flavours in the dough. Accordingly most bakers will aim for a fermentation temperature of around 25 °C.

As you will see in Chapter 11, it is usually best to avoid the formation of gluten when making pastries. One way to reduce gluten formation is to coat the starch granules with fat before adding any water. The proteins around the starch granules are thus isolated from the water and are unable to swell and interact – i.e. the first stage in gluten formation is prevented, or significantly reduced. The technique used in pastry making is to rub the fat into the flour before adding any liquids. Thus it should be apparent that, in bread making, where we want to ensure a good deal of gluten formation, the fat should not be rubbed into the flour, but should be mixed in along with the liquids.

Key points when making bread
- Use a strong flour (i.e. one with a high protein content)
- Do not rub the fat in too much so that it interferes with the hydration of the starch proteins
- Knead the dough thoroughly
- Allow plenty of time for the dough to rise
- Knock down the risen dough and allow it to rise again at least once before cooking

Why are these key points important?

Bread rises because the carbon dioxide generated by the yeast makes bubbles in the dough. For these bubbles to have enough strength not to burst, you need to create the rubbery gluten sheets from the protein in a strong flour. The kneading is essential as it stretches the protein molecules and promotes gluten formation.

Basic Bread Recipe

Ingredients
750 g Strong white (or Bread) flour
20 g lard (or other fat, if using butter 30 g)
2 teaspoons salt (20 g)
$1^1/_2$ teaspoons sugar (10 g)
420 ml tepid water
Yeast – either 15 g fresh yeast mixed with a teaspoon sugar, or a sachet of traditional dried yeast activated according to the instructions on the packet, or a sachet of "easy cook" dried yeast.

Method

Mix together all the dry ingredients in a bowl and rub in the fat until the mixture has a uniform texture. Add the yeast and most of the water. Mix together with a wooden spoon so that the ingredients stick together. Add more water so that a stiff, but still slightly sticky dough forms. The stiffness of the dough is very important. If the mixture is too dry the dough will be so stiff that the bread will not rise well; on the other hand a very wet dough will lead to a bread which rises too much and has a coarse texture. You can only get the 'right' consistency for the dough with practice. There is no single correct consistency for the dough – different people prefer different textures to their bread. However, with a little practice and experimentation, you should quickly learn the best consistency for your own taste. At the end of this chapter I describe some simple experiments you can try in your own kitchen to help you to determine this consistency for yourself.

The next step is to knead the dough on a lightly floured work surface. Kneading is absolutely essential; it is almost impossible to knead the dough too much. While a very acceptable bread can be achieved with little kneading, experiments do show that the more the dough is kneaded the better the quality of the final product. Again, at the end of this chapter I describe a simple experiment you can try in your own kitchen to help you to decide how much kneading give the results you like best.

When you are kneading bread dough the idea is to compress and stretch the dough repeatedly so as to form the gluten protein complex. The commonest and probably most effective technique is to form the dough into a ball and then to press down on the ball with the palms of your hands while leaning forward, so that the dough is pressed down and stretched. If you then roll your hands you can form the dough back into a ball and repeat the process.

As you knead the dough it will become less sticky, stiffer and more elastic. You should carry on kneading for as long as you can keep it up – it can be quite hard work! I have found that I need to knead dough for at least 3 minutes to achieve an acceptable product, but that kneading for about 20 minutes gives the best results. However, the time taken will depend on just how much effort you put into the kneading.

Once the dough has been kneaded, put it in a bowl and cover with cling film, or a tea towel, to prevent it drying out, and leave to rise in a warm place such as an airing cupboard. Ideally the temperature should be around 20 to 25 °C (it should be above 15 and below 30 °C). Once the dough has roughly doubled in size (about 1 hour) take it out of the bowl and gently knock it down, with your fist, to it's original size and knead again for a few minutes.

Shape the dough to fit in the tin you are going to use for the loaf (or loaves) – this recipe will make enough for one large tin (about 400 by 100 mm) or two small tins (about 200 by 100 mm). Put the dough in the tins, cover and leave to rise again (until the dough has risen to twice its initial volume). Heat your oven to about 250 °C and cook the dough for about 25 minutes. Test the bread by tapping the crust – a cooked loaf will sound hollow. Take the bread out of the oven and remove from the tin. The bread should come out easily if you turn the tin upside down and give the base of the tin a sharp tap. Stand the cooked bread on a wire rack to cool before serving.

There are three types of yeast you may see on sale. Fresh yeast, which you can buy from some bakers and health food shops, is really a block of compressed yeast (usually containing a little cornflour to help keep it dry). The overall water content in fresh yeast is around 70%, with most of the water being inside the yeast cells. This form of yeast must be stored in the refrigerator, otherwise it will quickly spoil. As soon as sugar is added to the block of yeast, it will almost immediately begin fermenting the sugar and will produce some water so turning the whole block into a bubbling liquid. This mixture can then be added to any recipe requiring yeast.

Dried, or active dried yeast, is a product that is not often seen these days. In this form most of the water has been removed from the yeast cells, thus preventing them from carrying out any fermentation. On the addition of warm water to granules of dried yeast, many of the yeast cells are able to absorb enough water to come back to life and begin fermentation. However, a good proportion of the cells die completely in the drying process and are never activated again. To use this form of yeast you have first to add warm water to the dried granules and then add some sugar and leave until a vigorous fermentation has started before adding to the dough.

The third type of yeast is a recent innovation. As microbiology has improved, so better methods of drying yeast have been discovered, so that today it is possible to dry yeast cells to a moisture content of around 20% and then encapsulate them in a special emulsion, which contains nutrients for the yeast cells when they are re-hydrated. This form of yeast known usually as "easy bake yeast" needs no special treatment to activate and can simply be added along with all the other dry ingredients. Certainly, this form of yeast is very easy to use and more or less guarantees success.

Why the recipe works

The ingredients

Flour

Using the right type of flour is essential. The protein content must be high enough to allow gluten sheets to develop during the kneading stage (see panel).

Sugar

The sugar provides some 'food' for the yeast. There are some sugars in the flour that the yeasts can metabolise, so the sugar really provides a 'boost' for the yeast to allow it to get to work making carbon dioxide gas to make the dough rise quickly.

Salt

The salt acts to modify and control the action of the yeast. If too much salt is added the yeast will die before it has had the chance to make the bread rise. If there is too little salt and the dough is left to rise for a long time, the yeast may continue to multiply and give the bread a strong yeasty flavour. However, as I note in the panel about bread without salt later in this chapter, I have not found any significant deterioration in the taste of bread as the amount of salt is reduced to zero.

Fat

The fat is added to slow down the staling process in the bread. Basically, the staling process is a rearrangement of the starch molecules that involves water becoming tightly bound to the starch – a detailed description is given in the panel on staling. The addition of fat slows down the rate at which water can associate with the starch in the cooked bread and so helps to keep the bread fresh longer.

Why Bread Goes Stale

Flour is made up from small starch granules in which the molecules (especially the more linear amylose molecules) are highly ordered in small crystals. These crystals are formed as the starch molecules are synthesised by the growing plant. However, when we cook starch the structure is changed, and all the crystals are melted and destroyed. After cooling, the molecules in the starch granules can start to form new crystals. However, they aren't able to get back into the same arrangement as they had when the plant first made them, so they adopt a different crystal form. In this new crystal form, the starch molecules need to associate with lots of water before they can crystallize. It is this recrystallization which we recognise as staling. As more new starch crystals grow so they take more of the "free" water from the bread and it seems to "dry out". We can slow down the rate of crystallization by adding fats that tend to prevent the water molecules being absorbed by the starch granules.

The reason why the French buy fresh bread twice a day is that the bread stales quickly because they use no fat in it. The lack of fat is also largely responsible for the distinctive flavour. Nowadays the French often complain that their bread isn't as good as it used to be; this is because the bakers now use a range of other (non fat based) anti staling agents, and the French are not used to fresh bread! An interesting aside is to note that the rate of crystallization of starch is maximum at about 4°C, thus your bread will go stale most quickly if you store it in the refrigerator.

The Instructions

Kneading

Kneading is necessary to produce the gluten from the proteins around the starch granules. The gluten forms sheets that have similar properties to rubber. When bubbles of carbon dioxide form in these sheets and expand as the yeast continues to work, so the bubbles are blown up just like balloons. If the gluten sheets are too weak (too little kneading) then these bubbles will burst very easily and the bread won't rise properly.

Leaving to rise

The rising is the stage where the yeast 'works' to produce carbon-dioxide gas that inflates the gluten 'balloons'. The more of these small bubbles that form and the larger they grow the more the bread rises and the lighter the final product will be. Since the yeast 'works' best at temperatures around 20 to 25 °C, it is best (and quickest) to keep the dough warm during the rising. The rising itself stretches the dough and so the dough "kneads" all by itself and helps to make yet more gluten. Rising can be slowed down by reducing the temperature, one result is that the yeast behaves differently and new flavours develop. Many people find these flavours attractive, and some bakers rise their doughs in a fridge overnight to produce these flavours in breads sold as "overnight loaves".

Knocking down and rising again

The reason you knock the dough down and let it rise again is to try to make more, and smaller, bubbles in the final bread, so giving it a finer, but still light texture. When a bubble is closed up by 'knocking down' it will often form several smaller bubbles as the gluten sticks to itself. These smaller bubbles will then be blown up again as more carbon-dioxide is generated during the second rising.

knead the dough for a second time. The dough is left to rise for a second time, before the heater is turned up to a much higher temperature and the bread is cooked. The machine then turns itself off and your bread is ready to take out of the tin. These machines can be programmed to start up at an appropriate time so that the bread will be ready exactly when you want it.

If all this seems too good to be true, there are a few drawbacks. First, the bread is cooked at a lower temperature than in a normal oven which leads to a softer loaf. Secondly, since the whole machine is much smaller there is much more steam around during cooking leading to a texture more like the soft sliced breads some people buy from supermarkets! Thirdly, although you can vary the recipe to suit your own taste, the need to comply with the rising times selected by the machine's own programmes, means that if the bread is to rise sufficiently, you will generally have to add more sugar than in other recipes leading to a sweeter tasting loaf than many people prefer. Nonetheless, the bread made by these machines is very acceptable, and they certainly do take all the hard work out of kneading dough. Bread machines can make up to 3/4 kg of bread in about 2 hours.

What could go wrong when cooking bread and what to do about it

Problem	Cause	Solution
The dough does not rise	You have forgotten to knead the dough! Or The yeast is not acting properly	Make sure the dough was kneaded. Either start over again, or add some more active, yeast. If using dried yeast throw the whole packet away and buy a new supply. If using fresh yeast, make sure the yeast and sugar mixture has started bubbling away before adding to the dough.
Bread has 'crumbly' texture	The gluten sheets have not formed. Either: Insufficient Kneading Or: Insufficient protein in the flour	Knead longer next time Use a strong bread flour (a protein content of at least 12.5 g per 100 g)
Bread very coarse	The bubbles in the dough became too large. Either: the gluten sheets were not stiff enough because the dough was not kneaded sufficiently Or: the dough was too soft Or: the dough was left too long in the rising stage	Knead longer next time Use less water (or more flour) next time Do not allow the dough to rise to more than about 4 times its original volume.

Problem	Cause	Solution
Bread has very dense texture	The bubbles in the dough were too small	
	Either: the dough was too stiff	Add more water next time
	Or: the dough was not left long enough to rise	Leave longer to rise
Texture is dense at bottom but ok at top	This usually happens if the dough is only allowed to rise once, or if the rising time is too short	Rise the dough twice knocking down between rises. Let the dough at least double in volume on each rise.

Bread without salt

When I was young we lived in North London, and we had our bread delivered daily by the baker – in those days this was quite commonplace. One day the bread tasted dreadful and we simply could not eat it, so we used up some stale bread as toast that day.

The next day the baker came round to apologise and told us the problem was that his apprentice had forgotten to put salt in the dough mixture.

So I have known since I was very young that salt is essential in bread making. Recently, while thinking about writing about bread in this book, I decided to experiment to see how much salt was really needed in bread. So I made some breads with less and less salt in the dough to see when it started to taste so disgusting that I could no longer eat it as happened with the bread delivered to my parents' home all those years ago.

To my surprise, even when I added no salt at all to the dough, the bread tasted fine. So now, more than 30 years after the event, I realise the baker was covering something else up with his story of the salt.

I've given the matter some careful thought and I think that what probably happened is that somebody added something else instead of the salt. There would have been a variety of white powders in a bakery that could have been mistaken for salt. The most likely candidate to make bread taste so awful no one could possibly eat any is some sort of soap powder! So I guess that a disgruntled employee probably deliberately spoilt a batch of dough by swapping the salt with the soap powder. In any event it was all so long ago there is nothing I can do about getting any recompense now.

Recipe variations

Whole meal (Brown) bread and Granary Bread

Ingredients

750 g Strong wholemeal (brown) flour (for Granary Bread use 800 g of a granary bread flour or a flour to which some malted grains have been added)
20 g lard (or other fat, if using butter 30 g)
2 teaspoons salt (20 g)
450 ml tepid water

Yeast – either 15 g fresh yeast mixed with a teaspoon sugar, or a sachet of traditional dried yeast activated according to the instructions on the packet, or a sachet of "easy cook" dried yeast.
(Nb this recipe does not call for any added sugar as there is enough sugar in the flour to feed the yeast. However, a little sugar may be added for a sweeter flavoured bread if wished.)

Method

Mix together all the dry ingredients in a bowl and rub in the fat until the mixture has a uniform texture. Add the yeast and most of the water. Mix together with a wooden spoon so that the ingredients stick together. Add more water so that a stiff, but still slightly sticky dough forms. As described in the white bread recipe, the stiffness of the dough is very important.

Knead the dough on a flour covered work surface for 5 to 10 minutes and then put it, covered with cling film, or a tea towel, to prevent it drying out, in a warm place such as an airing cupboard. Once the dough has roughly doubled in size (about 1 hour) knock it down to it's original size and knead again for a few minutes.

Shape the dough to fit in the tin you are going to use for the loaf and leave to rise again. Heat your oven to about 250 °C and cook the dough for about 25 minutes. Test the bread by tapping the crust – a cooked loaf will sound hollow. Take the bread out of the oven and remove from the tin. The bread should come out easily if you turn the tin upside down and give the base of the tin a sharp tap. Stand the cooked bread on a wire rack to cool before serving.

Breads and rolls with different shapes

Follow the ingredients and instructions for either white, whole meal or granary bread. After the first rising stage, knead the dough and then divide it into pieces which have about half the size of the finished rolls you want. It is a good idea to weigh these pieces of dough so that all the rolls come out more or less the same size. A rough guide is that small dinner rolls should weigh about 50 to 100 g each, while rolls for filling as lunch are best at about 150 to 200 g each.

Carefully roll each piece between your hands to form it into a ball shape, place these balls on a baking sheet leaving plenty of space between them to allow them to expand as they rise. Leave to rise for an hour or so, before baking in a hot oven for 10 to 15 minutes. If you want a particularly crusty finish brush a little salt water over the rolls before baking. The water evaporates leaving behind a thin layer of salt on the surface of the rolls. Salt is strongly hygroscopic (it just loves to suck up water) so it keeps the surface dry and ensures a good crust.

The actual shape of the rolls will depend on how you shape them before the final rise. If you leave them as ball shaped, the finished rolls will be roughly spherical – such rolls look very attractive when served on a side plate with dinner. If you flatten the ball shaped pieces of dough before the second rising, then you will obtain flatter, bap shaped, rolls ideal for filling with cheese, etc. for packed lunches.

You can also vary the ingredients to make sweeter tasting rolls and to obtain a softer texture in the finished breads. To make the bread have a sweeter taste add extra sugar, up to 50 g to the ingredients. To obtain a softer texture add some dried milk powder (up to 50 g).

Naan

Naan is a traditional Indian bread commonly served in Indian restaurants in the UK. Naan is cooked in a tandoor oven – it is not a simple matter to reproduce the effect of cooking in such a oven in the domestic kitchen. However, a combination of cooking in the oven at the highest possible temperature and then finishing under a very hot grill produces an acceptable version. If you have a charcoal, or gas fired barbecue, it can also be used for this sort of bread, often with greater success.

Ingredients
300 g Ata or Chapati flour (If this is not available any Strong flour will suffice – but will give a less authentic flavour)
30 ml vegetable oil
$^1/_2$ teaspoon salt (5 g)
1 teaspoon sugar (10 g)
100 ml plain (unset) yoghurt
100 ml milk
Yeast – either 10 g fresh yeast mixed with a teaspoon sugar, or a sachet of traditional dried yeast activated according to the instructions on the packet, or a sachet of "easy cook" dried yeast.

Method

Warm the milk and yoghurt to body heat and then combine all the ingredients together in a bowl and mix well to form a soft dough. Knead this dough for at least 5 minutes and then set aside in a warm place to rise (for about 1 hour). Preheat your oven to its highest temperature and put a heavy baking sheet in the oven. Knock down the risen dough and divide it into 4 pieces, use a rolling pin to shape these into ovals. Leave to rise for about 10 to 15 minutes and then lightly brush with melted butter. Switch your grill on at full heat. Put the naan onto the hot baking sheet in the oven and bake for 3 to 5 minutes – the naan should puff up during this stage. Remove the baking sheet and place it under the hot grill (as close as possible) for a further 3 minutes so that the surface of the naan is well browned – almost burnt. Keep the naan warm by wrapping in a tea towel and serve with the meal.

Unleavened Breads

There is a wide variety of unleavened breads (breads made with no raising agent such as yeast) cooked traditionally around the world. Probably the best known of these are the Indian chapatis and the Mexican tortillas. Although these come from very different cultures they still have a good deal in common. The basic ingredients are very similar – both consist of just flour and water. Chapatis use wheat flour while tortillas are traditionally made with flour from ground maize. The method in either case is to mix the flour with enough water to form a soft dough which is then kneaded, rested for a few minutes and then kneaded again. The dough is then formed into thin flat pancakes – experts do this with their hands throwing the dough from hand to hand while letting it spin. However, unless you wish to spend years practising the technique, it is much simpler to roll the dough out into thin sheets; although these are not likely to be as thin as those spun by hand. The breads are then cooked by placing directly on a hot heavy plate for a short time (typically only 30 seconds each side). During the cooking the chapatis, or tortillas will puff up and some black spots may form where the dough burns a little on the hot plate.

The main differences between chapatis and tortillas lie in the flour and the cooking temperature. Chapatis are made with a whole wheat flour and are typically cooked over a moderate heat using an iron plate, or heavy based frying pan on a medium burner and traditionally finished by placing them directly on hot charcoal for a few seconds. Tortillas are made from a yellow maize flour and are usually cooked at a higher temperature again using an iron plate or frying pan this time on the maximum heat available.

Pizzas

Pizzas are really quite simple to make at home. The basic dough is made exactly as that for white bread, although if it is available you should use a flour that has a high Durum wheat content – such flours are often available from specialist food shops and are often marketed as pasta or pizza flours. Follow the recipe for white bread (but use smaller quantities as below).

Ingredients
225 g Strong white (or Bread) flour
5 g lard (or other fat, if using butter 10 g)
$^1/_2$ teaspoon salt (5 g)
$^1/_2$ teaspoon sugar (5 g)
125 ml tepid water
Yeast – either 10 g fresh yeast mixed with a teaspoon sugar, or a sachet of traditional dried yeast activated according to the instructions on the packet, or a sachet of "easy cook" dried yeast.
Toppings of your choice.

Method

Form the dough as for the white bread and then leave to rise for 1 hour before rolling into thin round pizza bases. The quantity above is about right for two 25 cm diameter pizza bases. However, some people prefer the crust on their pizzas to be thicker in which case you should either use more dough, or make smaller pizzas.

Preheat the oven to its highest temperature while you put the desired filling on the pizzas. You should start by lightly coating the bases with a mixture of tomato puree warm water and oregano, the mixture should be fairly thick – you will need to use about one part water to two parts puree. The oregano should be added to taste. Simply take a ladle of the mixture and pour it in the middle of the pizza and then spread it around with the back of the ladle in a spiral pattern. You should stop just short of the edge of the dough. Next pile some grated cheese (a mixture of cheddar and mozzarella works well) over the tomato puree to a depth of about 1 cm. Now add any other fillings of your choice – mushrooms, ham, peppers, etc. and then a second layer of cheese. Finish the top with a little more oregano, some ground black pepper and a sprinkling of olive oil. Place the pizzas in the oven and cook for 15 to 20 minutes until the cheese has melted and the crust at the edges has browned. Use a fish slice to transfer the pizzas to plates and serve.

Some experiments to try at home

An experiment to see how much gluten you form in your dough

The gluten is mostly insoluble in water, so make a dough with just flour and water, weigh it, and then put it in a sieve and run water from the tap over it. At first quite a lot of the flour in the dough will be dissolved in the water and will run through the sieve, after a while you will be left with a semi-solid, sticky mess in the sieve, squeeze this to remove any water and weigh it to see what proportion of the dough you converted into gluten. Repeat the experiment kneading for different times to see how the amount of gluten formed increases as you knead the dough.

An experiment to determine the "ideal" consistency of bread doughs

Some cooks prefer to use very soft doughs and others like bread from very stiff doughs. For this experiment, you will need to make several different doughs adding increasing amounts of water to each dough. Use the basic bread recipe above with the ingredients given below

750 g Strong white (or Bread) flour
20 g lard (or other fat, if using butter 30 g)
2 teaspoons salt (20 g)
$1^1/_2$ teaspoons sugar (10 g)
350 ml tepid water
Yeast – either 15 g fresh yeast mixed with a teaspoon sugar, or a sachet of traditional dried yeast activated according to the instructions on the packet, or a sachet of "easy cook" dried yeast.

Begin by mixing all the ingredients. Divide into 8 pieces. To the first piece add no water, and to the next add 10 ml more water, to the next add 20 ml extra water and so on adding 70 ml extra water to the last piece. Knead all the pieces for the same time.

Next measure the "elasticity" of the doughs. For each dough roll out a piece 5mm thick and cut it into a strip 1 cm wide and 10 cm long. Pull this strip until it is 15 cm long (noting how hard you need to pull) and then leave it to relax for 10 minutes – measure how long it is. The more the dough remains stretched the less elastic it is.

Next measure the "stiffness" of the doughs. Roll a piece of each dough into a ball about 1cm in diameter. Place a 200 g weight on top of the dough ball and see how far it is compressed. The more compressed the dough the less stiff it is.

Finally make small rolls from all the doughs and cook them – taking care that you know which bun comes from which dough. Eat the buns and decide which you like the best. Now you can modify the amount of liquid in the basic recipe to your own taste.

An experiment to see how much you need to knead bread dough

Make up 8 pieces of dough according to the same recipe, using the same amount of water for each dough. Simply knead each piece for different times – again note how elastic and how stiff they all are (this will help you get the same consistency next time). Cook rolls from each dough and choose the one you like best to guide you as to how much you should knead your bread in future.

Sauces

Introduction

Sauces are integral to many dishes, from the simple vinaigrette on a salad to the complex rich Mexican moles. Sauces are used to provide both flavours and to enhance textures of nearly everything we eat.

There are two separate steps in making sauces; obtaining the required flavour and ensuring the sauce has the right 'thickness' and 'mouth-feel'. The flavours come from many different sources; in gravies the flavour comes from the Maillard reactions that have occurred during the roasting process, while in other sauces the flavour is intrinsic to the ingredient, for example the cheese sauce in Chicken Parmesan.

In this chapter we will concentrate on the thickening processes used in sauces and only briefly touch on the flavouring of these sauces as appropriate. However, we will address the issues involved in preparing stocks. Stocks are used as the basis of most savoury sauces and hence deserve a little consideration.

Thickening Mechanisms

There is a fairly good understanding of the various processes involved in the thickening of sauces in molecular terms. However, some of the concepts involve quite advanced areas of physics and chemistry, which can easily leave the layman quite baffled. The most difficult area is to understand what makes a sauce (or any other liquid) 'thick' or 'thin'. The first problem lies with the language we use to describe the 'texture' of a sauce. The term 'thick', when applied to a sauce covers several different properties that are controlled by different underlying molecular mechanisms. The first property is that of 'viscosity'.

The viscosity of liquid can be defined as the ratio of the rate at which it flows through a pipe, to the pressure applied to the liquid to make it flow. A 'thick' sauce will need a high pressure to push it through a small pipe, so it will have a high viscosity. In the panel below I have discussed viscosity and how it can be controlled in a little more detail. However, the properties of sauces can be rather more complicated. We often want a sauce to pour easily from its container on the table, but to coat the food over which it is poured and not to run all over the plate. In other words we want the viscosity of the sauce to change as it is poured. We want the sauce to have a low viscosity when it is being poured and a high viscosity when it is stationary on the plate. Liquids that display such properties are termed "shear thinning" or thixotropic. Generally, solutions of very long molecules in a low viscosity liquid, such as water, will be shear thinning.

Two main methods are used to thicken sauces. In the first, starch granules are swollen in hot water and in the second proteins are crosslinked to form large networks which can in turn form gels. Both methods are used in all sorts of sauces, both sweet and savoury. It is worth learning a little about the general scientific principles behind each method before we go to discuss details of the preparation methods of different sauces. So in the following section we will see how water (which is after all the main ingredient in all sauces) is made 'thicker' by the various methods used in cookery.

'Thickness' and Viscosity

Where most people talk about the 'thickness' of a sauce, the scientist will refer to its 'viscosity'. Viscosity is a measurable property of any liquid, and describes how fast a liquid will flow under a particular pressure. Imagine you had a tank of the liquid in the roof of your house and that the liquid flows down through some pipes to fill up your bath. The 'head' of the liquid above the bath provides the pressure that is pushing the liquid through the pipes. Then, the higher the viscosity of the liquid, the longer it will take to fill your bath. Strictly the pressure is given by the product of the density of the liquid, the height of the tank above the bath and a constant, g, called 'the acceleration due to gravity'.

A thick sauce is a viscous sauce. To understand how to make a thick sauce it helps to appreciate the reasons why some liquids are more viscous than others. When a liquid flows, the molecules inside the liquid have to move past one another. It is more difficult to push a large molecule out of the way and it is more difficult to slide easily past an irregularly shaped molecule. Thus, in general, liquids made up from larger and more irregular shaped molecules have higher viscosities.

Of course, sauces are mostly water, so we have to find ways to make water more viscous if we are to thicken a sauce. If we dissolve some large molecules in the water, then they can increase the viscosity and hence thicken the sauce. The larger the molecules the fewer we will need to get a truly viscous sauce. You can increase the viscosity of water by adding small molecules, such as sugar. In fact all cooks already know, but may not have consciously realised, that very concentrated sugar solutions are very viscous – think of 'Golden Syrup', Treacle and Honey – all concentrated sugar solutions and all quite viscous and thick. The problem with using sugar as a thickening agent is that to achieve sufficient thickening you would have to add so much sugar that the sauce would be overwhelmingly sweet. As explained in Chapter 3, small molecules are normally volatile and carry flavours with them as they evaporate into the nasal passages, so any sauce thickened solely by the addition of a very large proportion of small molecules will possess a strong, and usually overpowering, flavour.

What is needed to achieve thickening without introducing overpowering flavours is to add very large molecules that can, due to their size, thicken a sauce in quite small amounts; and due to their large size, do not impart any particular flavours. As we have already seen there are two main types of large molecules used in foods – starches and proteins. Not surprisingly, both are used as thickening agents in a variety of different sauces. However, there are some further subtleties involved. The starch comes in the form of 'granules' that swell up in hot water – so it can be the swollen granules themselves rather than the starch inside them that thickens the sauce. Many proteins are a lot smaller than starch molecules and it can be helpful to make them into larger molecules by heating them until they react together and form a 'network' molecule that can be many hundreds of times larger than a single protein. Such large molecular aggregates can have as much as a thousand fold greater effect on the viscosity than the separate proteins before they 'coagulated'.

Starch Based Sauces

As we have already seen in the previous chapter and in Chapter 2, starch is formed in small granules in many vegetables, and seeds. There are two main molecules (amylose and amylopectin) involved in making starch granules. These long molecules are made up from long strings of small sugar molecules joined together. In amylose the sugars are joined to make linear strings, while in amylopectin, the sugars join in a more complex fashion to make branched molecules (see Chapter 2 for a detailed description of these molecules). Although made up from sugar molecules, these long, starch, molecules do not taste at all sweet. The sweetness sensors on our tongues do not react to such large molecules.

On heating above about 70 °C, the starches amylose and amylopectin start to become soluble in water and the granules begin to absorb large amounts of water. As more water flows into the granules so they expand; in extreme cases they can expand to 100 times their original volume. These expanded granules provide a good deal of thickening by increasing the viscosity of the liquid.

To see how this increase in viscosity occurs you can make a simple experiment with a washing up bowl, 10 to 15 balloons and some water. The balloons act as large scale models for the starch granules. Begin by putting just a little water into each balloon so that it is still quite limp to represent a starch granule before it is swollen. Now put all the balloons in to the washing up bowl and fill it with more water. If you stir the water in the washing up bowl it is easy to move your hand through the water – so you could say the liquid has a low viscosity, or the sauce is thin. Next, take the balloons and put a lot more water into them, so that they increase in size to about 5 times their original diameters. Now put the balloons back into the washing up bowl with enough water to fill the bowl and try stirring again. Now it is very difficult to stir the water/balloon mixture as you have to deform the balloons to be able to move between them. You could say you have a high viscosity liquid (or thick sauce) in the bowl.

However, this is not the end of the story. In the sauce, it is possible for some of the starch molecules to burst out of the granules (this will happen at sufficiently large expansions of the granules – i.e. at sufficiently high temperatures). Once this happens, the "free" starch contributes to the general thickening but also, by forming a sort of entangled network of long molecules that penetrates throughout the sauce, they impart a shear thinning property to the sauce, which will allow it to be poured, while it will also not flow too much on the plate.

Starches are used widely to thicken gravies, most commonly as cornflour, or wheat flour. Another major use of starch to thicken a sauce is in some custards; the best custards are thickened with egg proteins, however, this can be a time consuming and expensive process, so most commercial custards (and all 'instant' custards) use cornflour as the thickening agent.

Mr Bird's Custard
Mr Bird was a pharmacist whose wife had an allergy to eggs, but loved custard. In order to allow her to enjoy custard, he invented a new way of making custard using cornflour as the thickening agent, rather than the egg yolks that had always been used until then. The recipe was simple, milk, sugar and cornflour were heated together along with some flavourings (vanilla, etc.) and colourings (amaretto, etc.) until they thickened and formed a sauce that was not too dissimilar to the traditional egg based custard. Mr Bird developed the recipe and sold the ingredients in his chemists shop. Soon, many people came to like the ease and cheapness of his custard and he set up a company to manufacture the product as a simple powder that was soon sold around the world as Bird's Custard. The product is still available in more or less its original form today from any supermarket anywhere in the world.

Protein Thickened sauces

As we have seen in Chapter 2, proteins are long molecules made up from linear strings of amino acids joined together. In nature these proteins have well defined shapes that define their biological functions. However, when heated, the proteins change their shapes in a process known as denaturation (see Chapter 2 for details). Once the proteins have denatured they tend to expand and spread out in any surrounding water. As the temperature is raised proteins will come together and react with one another forming bridges between adjacent molecules, thus building up large networks that can act to thicken a sauce.

An example of a protein thickened sauce is an egg custard. The proteins in the egg yolks are first denatured at a temperature of around 40°C, and then react with one another to form a network at temperatures above about 70°C. If too much of this networking should occur, the custard will become too thick and lumps start to form, or, in extreme circumstances, it may turn into scrambled eggs! A good guide is to make sure the temperature of your custard never gets above 80°C.

Texture and 'Mouth Feel'

While it is the viscosity, or thickness, that largely determines the texture of a sauce, there are several other factors that can be important. Food Technologists often lump these factors under the heading of 'mouth-feel', a rather ugly term used to describe the sensations of eating food. 'Mouth-feel' includes all aspects of the texture of the food as well as some aspects of flavour (such as how long the flavour and taste persist in the mouth). As far as sauces are concerned you may be interested in 'smoothness', 'lumpiness', creaminess, acidity or 'sharpness', as well as the persistence of the texture and flavour. It is very difficult to separate these different qualities and then pick out the aspects of making a sauce that affect them.

However, a few simple guidelines are available. The sensation of 'creaminess' comes largely from the tendency of thick creamy foods to coat the mouth and not to dissolve very quickly so providing a taste sensation that persists for some time. That is to say, creaminess comes from a combination of the viscosity (thickness) and the solubility of the fats in the sauce. Control of viscosity has been discussed above, so we shall concentrate here on the incorporation of insoluble fatty material to improve the 'creaminess' of a sauce. Creaminess can be achieved most easily by the addition of some thick cream; the fat in the cream is at best poorly soluble in the mouth and being very thick it can readily coat the inside of the mouth. Indeed, this 'creamy' texture is the reason why many recipes call for the finishing of a sauce with the addition of a little cream at the last minute.

However, not all 'creamy' tasting sauces have cream added to them. Other methods can be used to incorporate some small droplets of insoluble fats (or other substances such as starch granules that have been treated to prevent swelling) into a sauce. In scientific terms a suspension of fatty droplets in another liquid is called a colloid and there is a whole branch of physical chemistry devoted to the study of colloids. Essentially, to make a colloid you need to form very finely dispersed drops of the fat and coat these with some 'surfactant' molecules that stabilise them in the water based medium. Surfactants are molecules one end of which likes to be in the fatty environment (and does not like to be in the watery environment of the bulk of the sauce) and the other end of which likes to be in the watery environment of the bulk and does not like to be in the fatty droplets. There are many examples of such molecules that can be used in making colloidal sauces – probably the most important are the 'lipids' found in egg yolks.

Adding egg yolks to a mixture of oils and water and then vigorously stirring the mixture will create a reasonably stable colloidal suspension of oil drops in the water and (providing the stirring is sufficiently vigorous) make a thick and creamy sauce. Probably the best examples of such sauces are the many varieties of mayonnaise.

Stocks

Most savoury sauces are based on some stock or other, the purpose of the stock being to provide the base flavour for the sauce. Years ago, when cooks had plenty of time, stocks were normally prepared well in advance as and when ingredients were available – often being made from all sorts of left overs and scraps. Today, most cooks have far too little time to make up stocks which have no immediate use, and the use of pre-prepared stocks, and stock tablets, has become commonplace. While these proprietary stocks can often give a good result, nothing beats a sauce made from a really well prepared stock. The main reason why home made stocks seem to give better flavours probably comes from the fact that there is no need to add preservatives to home made stocks. If you look at the list of ingredients in a stock tablet or most proprietary liquid stocks you will find the first listed ingredient (i.e. the one there is most of) will be salt. The salt is not there for flavour – indeed there is so much of it they sometimes have to add other things to disguise just how much there is; no, the salt is there to act as a preservative – if there is enough salt then bacteria and mould can't grow and the stock will have a long shelf life.

If you do decide to make your own stocks there are two ways you can keep them safely for quite long times (several weeks at least). First, you can concentrate them by boiling them down to a thick dark paste which you can dilute later. Such pastes (or marmites) will keep fairly well in a fridge for a week or two. Of course you should make sure you boil them up again when you do get around to using them. Your second option to keep stocks is to freeze them – you can of course concentrate them before freezing if you are short of space in the freezer. I know some people who fill ice cube trays with concentrated stocks.

To make a good stock you need to create and extract as much flavour as you can from your base ingredients. So when making meat stocks you need to ensure that the Maillard (browning) reactions take place to create plenty of flavour molecules (see Chapters 2, 4 and 6 for more details) and imbue the stock with a strong meaty taste. For a vegetable stock you will want to try to ensure that as much of the character of the vegetables as possible is incorporated into the stock. Whatever kind of stock you are making you should also think about its colour – for white sauces you will need a clear stock, while for brown sauces a dark stock can be an advantage.

The recipes below describe how to prepare basic vegetable and meat stocks.

Vegetable Stock

Key Points to consider when preparing vegetable stocks

- Always include some onions and carrots in the stock. Onions, when cooked, provide flavour enhancing molecules – these act a bit like monosodium glutamate (MSG) and help to accentuate other flavours present in the mouth through the Umami sensation (see Chapter 3 for details). Carrots give good colour and can add thickness to a sauce.
- 'Sweat' the vegetables to soften them and to extract flavour before adding any extra liquid.
- Control the colour of the sauce by allowing the vegetables to brown as much, or as little as you wish.

Ingredients (to make about 1.5 litres of stock)
250 g carrot   (one large carrot)
300 g onions   (two medium onions)
200 g mushrooms
250 g leek   (one leek)   optional
250 g parsnip (one medium sized parsnip)   optional
2 litres   water

For a thick sauce also add
250 g potatoes   (two medium potatoes)
NB you can add any other vegetables you wish and you may vary the proportions to suit availability and personal choice.

Method

Clean and chop the vegetables into small (about 1 cm) sized pieces. Put them in a thick-bottomed enamel or stainless steel pan on a medium low heat. Do not add any water or fat or oil. Put the onions and leeks in first followed by the other ingredients finishing with the mushrooms. Keep the pan covered tightly and stir the vegetables regularly until they have softened and collapsed down to about half their original volume. Now keep a careful eye on the vegetables as they begin to brown on the bottom of the pan. You should decide how dark you want the stock to be and allow sufficient browning.

For a thick dark sauce, allow a layer of browned vegetable matter to stick to the bottom of the pan and then turn up the heat and let the colour deepen to a dark chocolate colour before adding a little boiling water. With a wooden spatula, scrape the browned vegetables off the bottom of the pan and add more water as needed. Add more water until it covers the vegetables and then add about half as much again – this should be about 2 litres of water. Continue to cook the covered vegetables on a low heat for at least another 40 minutes and then strain and scrape the mixture through a sieve; after passing all the vegetables through the sieve you should have only about a small cup full of solid matter left over.

For a thin dark sauce do exactly as above, but do not force the vegetables through the sieve at the end of the cooking, instead only keep the liquid that the vegetables were cooked in.

For a thick light sauce add the boiling water directly to the cooking vegetables as soon as they start to brown and simmer for at least an hour. Only when the vegetables are really soft and starting to break up should you strain and scrape the mixture through a sieve; after passing all the vegetables through the sieve you should have only about a small cup full of solids left over.

For a thin light sauce do exactly as above, but do not force the vegetables through the sieve at the end of the cooking, instead only keep the liquid that the vegetables were cooked in.

Concentrating the stock

If you want a more powerful flavour from your stock, or if you want to keep some stock for future use, you can concentrate it readily by leaving it boiling on a medium heat until the volume is reduced from about 1.5 litres to about 100 ml. Once reduced the stock will keep fairly well in a fridge for a couple of weeks or several months if frozen. Of course, when you reduce a stock as described above, the flavour will change. The different flavour molecules in the stock will all evaporate, but at different rates, so the relative proportions of these molecules will change as the stock is reduced, leading to variations in the flavour.

What could go wrong when making vegetable stocks and what to do about it

Problem	Cause	Solution
Stock has bitter taste	The vegetables were burnt rather than browned.	There is no real solution; however, the bitter taste can be masked with a little sugar. Next time don't let the vegetables burn during the sweating stage.
Stock not dark enough	The vegetables were not sweated long enough at the beginning	Either reduce the stock to increase the depth of colour, or make a little browning by soaking some well browned, fried vegetables (onions, carrots) in a little water and add to give colour.
Stock not clear	You have removed too much starch from the vegetables.	If you need a clear stock you should avoid passing the vegetables through a sieve and removing starch from them. Also avoid using potatoes in the stockpot. You may be able to clarify the stock, see note after section on meat stocks later in this chapter.
Stock has little flavour	Either not enough vege-tables used, or vegetables not sweated long enough.	Reduce the stock to increase the intensity of flavour. Next time use a greater quantity of vegetables and/or sweat them for a longer time.

Meat Stock

Key Points

- Thoroughly brown all the meat and bones before adding any water.
- If possible use a previously prepared vegetable stock instead of using the vegetables as described below.

Ingredients (to make about 1.5 litres of stock)
About 150 g meat trimmings and any available bones

Either 2 litres of vegetable stock (see above) or:
250 g carrot   (one large carrot)
300 g onions   (two medium onions)
200 g mushrooms
250 g leek   (one leek)   optional
250 g parsnip   (one medium sized parsnip)   optional

Method

Cut the meat trimmings into small pieces, the smaller the pieces the more surface there will be where the Maillard (browning) reactions can occur and develop the flavour. Thus the more finely you chop these trimmings the more flavour you will achieve in the final stock.

If you are using vegetables, rather than a vegetable stock, cut them into small pieces and sweat them as described in the vegetable stock recipe above, for about 20 minutes while the meat is being browned. Then add about 3 litres of boiling water to the vegetables and allow to simmer uncovered for another 30 minutes, before straining.

Brown the trimmings; this is the most important step. Flavour is developed as the large protein molecules are broken down into smaller, volatile, 'flavour' molecules through the Maillard reactions (see Chapters 4 and 6 for further details of these reactions). To get the best flavour from the browned meat trimmings this browning should be at a temperature above about 140 °C (the Maillard reactions are rather too slow at lower temperatures) and below about 250 °C (at higher temperatures the Maillard reactions create different molecules that give food a bitter, burnt taste). You can achieve the browning either on the hob in a frying pan over a medium heat, or in the oven set at a temperature of around 200 °C.

If you use an oven, put the trimmings in a single layer well spaced apart, in a baking tin and roast for at least one hour, until the meat is very well browned. If possible turn the trimmings over once during the cooking.

If you use a frying pan use only a little oil, or fat, and cook a few pieces at a time, making sure the frying pan is never more than half covered, over a medium heat, turning the trimmings constantly until they are very well browned (a dark, chocolate colour with a good sheen).

If you have any bones available, chop, or split, these into pieces – breaking them open to allow access to the marrow will greatly enhance the flavour of any stock and brown them along with the trimmings.

Whichever method you use to brown the meat, the remaining steps are the same. Take the meat trimmings (and any bones) out of the pan they were cooked in and drain off any surplus fat. Boil up some of the vegetable stock and add to the pan (which should be on the hob at a medium heat). Allow the stock to boil as you scrape around the pan to remove all the reaction products that have stuck to it. Pour the stock into a saucepan and repeat the process with more stock until the pan in which the meat trimmings were cooked is quite clean. Cooks often refer to this washing of the browned products of the Maillard reactions from the hot pan to the stock or sauce as "deglazing". Add the rest of the stock to the saucepan along with the meat trimmings and any bones and simmer for at least 1 hour – the longer you can cook the stock the more flavour will be extracted from the meat trimmings – some recipes recommend simmering stocks for days! However, after an hour or so little further advantage is gained, unless you really are prepared to go on for several days.

While the stock is simmering, some scum will rise to the surface – this scum comes from some of the connective tissue in the meat and bones breaking down and will, unless it is removed, make the stock rather cloudy. So to obtain a clear stock you will need to skim off the scum as it rises to the surface. You should expect to do this skimming about every quarter hour as the stock cooks.

Finally, strain the stock off from the meat trimmings and bones and it is ready for use. You can reduce the stock as described for vegetable stocks above if you wish to keep it.

A note on clarifying stocks

Many sauces have a much better appearance when they are transparent, or at least translucent. If you want to make such a sauce it is absolutely essential that your basic stock itself is completely clear. While most vegetable stocks will be quite clear (unless you sieve the vegetables back into the stock at the end), meat stocks are often rather cloudy. This cloudiness develops from small aggregates (only a few thousandth of a millimetre in size) that form from the denatured proteins from the connective tissue that slowly dissolves during the long simmering times of the stock.

The traditional way to prevent stocks becoming cloudy is continually to skim the surface as the stock simmers – this method works because the protein aggregates rise to the surface to form a scum like layer – so if you skim them away as they rise the stock will remain clear.

Alternative and simpler methods are also available. You can prepare a stock without skimming it at all and then remove the small aggregates later (if you have the time). There are two main methods that may be used. For those with access to scientific equipment simple glass filters are easy to use and after filtration stocks become remarkably clear. However, if you can't get hold of suitable filtration equipment you can use the traditional finings methods used to clarify wines or beers. You simply add a 'fining' agent to the stock and leave it to stand. The 'fining agent' makes the small aggregates stick to one another to form much larger particles that sink to the bottom of the stock so that you can decant clear stock from the top. You can either buy commercial fining agents (such as isinglass) from anywhere that sells supplies for home made wine or beer, or you can use egg whites. Simply mix a little of the cold stock with the fining agent and then put it together with the rest of the cold stock in a tall bottle (such as a wine bottle). Keep the bottle cool in an upright position (ideally in a fridge or cellar) for a day or so until the finings have worked and you have a clear stock at the top of the bottle. Finally siphon off, or decant, the clear stock leaving behind the cloudy residue at the bottom of the bottle.

What could go wrong when making meat stocks and what to do about it

Problem	Cause	Solution
Stock has bitter taste	The meat was browned at too high a temperature.	There is no real solution; however, the bitter taste can be masked with a little sugar. Next time brown the meat at a lower temperature in the oven or on a lower heat on the hob.
Stock not dark enough	The meat trimmings were not browned enough	Reduce the stock until a suitable colour is achieved, or make a little browning from fried vegetables (onions, carrots) or meat trimmings and water and add to the light stock. Next time brown the meat trimmings more thoroughly.
	Not enough meat trimmings were used,	Next time use more trimmings and bones
	The meat trimmings and bones were not cut into small enough pieces	Next time cut the meat trimmings and bones into smaller pieces.
Stock not clear	The scum that rose to the surface during the simmering stage was not skimmed away.	The stock may sometimes be clarified by adding a little egg white and standing in a cool place overnight. The egg white should make the protein residues that are making the stock cloudy come together and sink to the bottom. You will need to decant the stock very carefully to prevent it going cloudy again.
Stock has little flavour	There was insufficient browning – either from too little meat and bones, or from not sufficient time for the browning to take place	Reduce the stock to increase the intensity of flavour. Next time use more meat trimmings, cut them up more finely and brown for longer at a higher temperature.

Recipes

Starch Based Sauces

Basic Principles and Key Points

The use of starches is probably the commonest and simplest way to thicken sauces. The most usual problem that may be encountered is that the starch granules stick together before they start to absorb water and expand to thicken the sauce. If lots of starch granules stick together in this way then they will form lumps in the sauce. The starch granules at the centre of such lumps have no access to the water and so do not become swollen up.

The key to making a good sauce is to ensure that the starch granules do not stick to one another before they swell up and thicken the sauce. Starch granules stick to each other when a little of the protein (or even starch) that is in and around the granules is partially released into the water surrounding the granule giving it a "sticky" surface. Then if two such "sticky" granules come together the protein molecules attached to them will interact to bind them together. As further granules join this initial pair a large aggregate can quickly be built up. The granules at the middle of the aggregate will not be surrounded by water, but rather by other starch granules, so when the water is heated only those starch granules around the edge of the aggregate can absorb water and swell. The aggregates then make the "lumps" in a "lumpy" sauce. The simplest way to prevent granules sticking together (and hence prevent any lumps appearing in your sauce) is to ensure that the granules are well dispersed before they are heated and swollen. Keeping the granules well dispersed prevents them from coming together and forming any aggregates, so that when they are heated all the granules are surrounded by water and are able to swell up to thicken the sauce uniformly.

There are several different ways in which the starch granules may be dispersed. The two main methods are to disperse the granules in *cold* water or to disperse them in a fat, or oil. In either case, the resulting suspension of starch granules (in fat, or water) is added to the sauce to be thickened and the whole mixture is kept well stirred as it is heated until the starch granules swell up and thicken the sauce.

In the recipes below, the basic white and brown sauces both use a method of thickening with wheat flour which is dispersed in fat (normally called a roux) while the gravies use the simpler approach of cornflour dispersed in cold water as the thickening agent. Of course, this distinction between using wheat flour dispersed in fat to make "sauces" and cornflour dispersed in water to make "gravies" is quite arbitrary and either method can be used in any situation. Most cooks seem to believe that the wheat flour method gives a better flavour and mouth feel, however, since this method is the more difficult to master, many cooks stick exclusively with the foolproof "cornflour" technique.

The reason for the different techniques for wheat flour and cornflour lies in the amount of protein in these two starches. Wheat flour contains much more protein than corn flour and is thus more prone to forming the unwanted 'lumps'. In cornflour there is so little protein coating the granules that they do not become sticky until they reach a quite high temperature. So it is possible to disperse them in cold water and add that dispersion directly to a hot sauce. When the dispersed cornflour granules are in the hot sauce they rapidly absorb water and the sauce thickens almost instantly. On the other hand the much larger amount of protein in wheat flour will absorb cold water to make the granules quite sticky, so they will stick together in cold water and form lumps. So the more complex roux method of dispersing the granules in a melted fat is used in preference to the dispersion in cold water which could still lead to some lumps.

Basic 'White sauce'

Key Points

- Make sure the flour is well dispersed in the melted butter.
- Use a lightly coloured stock.
- Remove from the heat once the butter is melted so the flour does not brown.

Ingredients
100 g Plain Flour
100 g butter
1 litre light stock (or other liquid such as milk, etc.)
Salt, pepper and other flavourings as required.

Method

Heat the stock, or other liquid, until it is simmering. In another saucepan that is large enough to cook the whole sauce, melt the butter and let it bubble for a couple of minutes, but do not allow it to darken at all. The pan should have a thick base so that it retains heat. If you use a thin pan it can cool quickly when the flour is added and the butter may solidify again. Remove the pan from the heat and add the flour all at once. Beat the flour into the melted butter to make a smooth 'roux'. This 'roux' should have the consistency of a smooth, creamy paste.

Add a little (about 100 ml) of the stock to the roux and beat it in vigorously. The stock should thicken immediately and the roux should remain as a smooth paste. Add a little more stock and beat that in. Repeat until the sauce starts to become liquid like and then slowly pour in the rest of the stock while stirring all the time. Once all the stock has been added return the pan to the heat and quickly bring to the boil stirring all the time to keep the starch granules from sticking together. Once the sauce is boiling leave it for two or three minutes before reducing to a simmer for about 10 minutes. This cooking process will ensure the sauce loses any 'floury' flavour as the starch mixes more evenly through the sauce. Finally season and add any other flavourings you may need for the particular dish you are making.

What could go wrong when making a white sauce and what to do about it

Problem	Cause	Solution
Sauce too thick	Too much starch was used to thicken the sauce.	Add more stock – you should find that the sauce starts to become liquid once you have added about half the stock to the roux. If it is still a paste at this stage a good idea is to remove about half of the paste before adding any more stock so as to ensure the final sauce is not too thick.
Sauce too thin	Not enough starch was used to thicken the sauce. The sauce was cooked too long	You may be able to prepare a little more roux and repeat the whole process. Sometimes, when a starch based sauce is allowed to cool and then reheated or when cold liquid is added after it has thickened, it can become thin. This happens because the starch molecules degrade at high temperatures and so lose their thickening ability. If this happens all you can do is start over again.
Sauce is 'lumpy'	The granules were not kept separate enough in the roux.	Next time beat the roux much more thoroughly. For now, you may be able to rescue the sauce by passing it through a sieve and then reheating it carefully.
Sauce is dark	The starch was heated before the stock was added, or a dark stock was added	Make sure you use a clear and light coloured stock. Make sure you remove the pan in which you are making the roux from the heat as soon as the fat has been melted and before you add any flour.

Recipe variation: Chicken Parmesan

Ingredients
4 Chicken Breasts (boned and skinned)
2 lemons
100 g butter
500 ml milk
50 ml cream
200 g grated Parmesan Cheese (preferably freshly grated from a block)
30 g plain flour
1 small onion
1 Bay leaf
5 peppercorns
About 2 cm piece of mace.

Method

If possible use a dish that can be used both on the hob and in the oven and which has a well fitting lid. If you don't have a suitable pan, don't worry; just fry the chicken in a frying pan and transfer to an ovenproof dish with a well fitting lid for the cooking in the oven, making sure to scrape all the "browned" juices from the frying pan into the new dish, so as not to lose any flavour.

Cut some of the lemon peel into thin strips (preferably use a "zester" for this purpose) and then extract the juice from the lemons. Brown the chicken breasts in 50 g of the butter and then add the lemon juice. Cover the pan and put in a moderate oven (160 to 180 °C) for 20 minutes.

Meanwhile, prepare the sauce. Simmer the peeled and sliced onion, peppercorns, and mace in the milk (using a good heavy-based saucepan) for about 5 minutes and then strain into a separate bowl removing the onion and spices. Melt the remaining 50 g of the butter in the pan, remove from the heat and add the flour. Beat to a smooth paste. Add a little of the milk (about 50 ml) to the roux and beat it in vigorously. The milk should thicken immediately and the roux should remain as a smooth paste, but now a little less stiff. Add a little more of the milk and beat that in. Repeat until the sauce starts to become liquid and then slowly pour in the rest of the milk while stirring all the time.

Once all the milk has been added return the pan to the heat and quickly bring to the boil stirring all the time to keep the starch granules from sticking together. Let the sauce boil for two or three minutes and then add three quarters of the grated parmesan cheese, leave to simmer for about 10 minutes, as the chicken is cooking.

Take the chicken breasts out of the oven, remove liquid from the pan and mix it into the sauce, bring the sauce back to the boil, turn off the heat and then stir in the cream. Pour all the sauce back over the chicken breasts and sprinkle the remaining Parmesan cheese over the top. Place the dish with the breasts, sauce and Parmesan cheese under a hot grill until the cheese browns. Serve with fresh pasta for a refreshing and simple meal.

Basic Brown Sauce

Key Points

- Make sure the flour is well dispersed in the melted fat.
- Use a dark coloured stock.
- Make sure the flour is very well browned before adding the stock.

Ingredients
100 g Plain Flour
100 g dripping, or other fat
1 litre dark stock (meat or vegetable according to preference)
A tin of tomatoes (Optional)
Salt, pepper and other flavourings as required.

Method

Heat the stock until it is simmering. In another saucepan that is large enough to cook the whole sauce, melt the fat and let it get very hot. The pan should have a thick base so that it does not deform at all under the high temperatures used. Turn down the heat and add the flour and cook it stirring all the time until it becomes very dark. Next add a little of the stock (about 100 ml) and stir it in well. As soon as you add this stock it will boil and clouds of steam will come up – so make sure you are well prepared and do make sure to use a wooden spoon with a long handle so that you avoid getting scalded. The stock will thicken and continue to darken. Add a little more stock and beat that in. Repeat until the sauce starts to become semi liquid then slowly pour in the rest of the stock while stirring all the time. Once all the stock has been added turn up the heat and bring back to the boil stirring all the time to keep the starch granules from sticking together. Once the sauce is boiling leave it for two or three minutes before reducing to a simmer for about 10 minutes. This cooking process will ensure the sauce loses any 'floury' flavour as the starch mixes more evenly through the sauce. Finally season and add any other flavourings, (including the tomatoes if used) to balance the sauce for the particular dish you are serving.

What could go wrong when making a brown sauce and what to do about it

Problem	Cause	Solution
Sauce too thick	Too much starch was used to thicken the sauce.	As above for white sauces
Sauce too thin	Not enough starch was used to thicken the sauce.	As above for white sauces
	The sauce was cooked too long	As above for white sauces
Sauce is 'lumpy'	The granules were not kept separate enough in the roux.	As above for white sauces
Sauce is not dark enough	The original stock was not dark, or the flour was not browned when making the roux.	Next time make sure the flour is well browned in the hot fat before adding any stock to make the roux.

Gravies

There are several variations possible when making gravies to accompany roast meats, etc. The basics are however always the same; the differences lie largely in the method of thickening the gravy. The basis of all gravies comes from the juices that have dripped from the meat during roasting and undergone Maillard reactions in the roasting pan, so developing the deep meaty flavours. Consequently, the most important steps are in making sure all this flavoursome material is incorporated into the gravy.

If you use flour as a thickening agent, you can make a roux with the flour and the fat from the joint in the roasting tin, follow the instructions for a brown sauce above. After removing the joint from the roasting tin, put it on a suitable hotplate and add the flour to the tin, stirring all the time to make a thin roux. Allow the flour to darken to provide some extra flavour as well as a wonderful deep brown colour in the finished gravy. Once the roux is ready, add the pre-heated stock a little at a time beating all the time and finally season to taste with pepper and a little salt.

If you use cornflour to thicken the gravy the process is simpler and quicker and there is no danger at all of lumps forming. However, the gravy will generally not be so dark in colour and will be a little less rich.

Key Points

- Make sure the meat is well browned
- Make sure you scrape up all the "browned" meat juices
- Use the best available stock

"Garlic" Sauce

Many years ago, I had my first experience of cooking with garlic. I was making a sauce to go with some steaks and following the recipe very carefully as this was something I had never done before. The recipe for the sauce called for a clove of garlic to be crushed and added at some stage. Although I had eaten dishes with garlic in restaurants, I had never used it myself before, so I had to go out specially to buy not only the garlic, but also a garlic press. Fortunately, even though there were no instructions with the garlic press it proved simple to operate.

As the sauce was cooking, the smell of garlic permeated through our house and I began to wonder whether my father, who, at that time, was a strict meat and two veg man who considered garlic quite outlandish, may have had a point. However, nothing ventured, nothing gained, so I persisted with the sauce.

At the meal we both agreed the sauce was very strongly flavoured so much so that neither of us could finish our portions. I was particularly concerned as the smell of the garlic continued to hang throughout the house for days afterwards and eventually realised I must have made some error in the cooking. It finally dawned on me weeks later that the problem was in the definition of a "clove" of garlic. Knowing no better, I had assumed the white bulb one bought from the supermarket was a clove. It was only when I asked an older and experienced colleague at work I discovered that was a bulb and a clove was just one of the small segments inside.

No wonder the sauce made with a whole bulb of garlic was so strong. In fact the sauce was not unpleasant, just extremely powerful, and the experience has in no way lessened my liking for garlic.

Basic Gravy for Roast Meats

Ingredients
1 litre stock (preferably a meat stock with a dark brown colour)
Scrapings and juices from roast
Salt and pepper to taste
50 g Flour, or cornflour to thicken
30 ml cold water

Method

Begin by straining the fat from the roasting pan and then use about half the preheated stock to deglaze the pan. Pour this stock into a suitable saucepan, rinse out the roasting pan with the remaining stock and add to the saucepan. In a separate dish, or cup, add about 30 ml cold water to the cornflour and stir to form a thick paste, stir in a little more cold water so that the paste becomes quite thin. Put the saucepan on to heat and season to taste. Add the cornflour mixture just as the stock starts to come to the boil. Stir the stock all the time as the cornflour is added and bring to the boil for a minute or so. The gravy will thicken as soon as the cold cornflour mixture is heated above about 70 °C, so it is essential to ensure you keep stirring vigorously to ensure an even dispersion of the cornflour.

Protein Based Sauces

Proteins, once denatured, will usually form crosslinks between themselves and quickly build up large networks that can readily thicken sauces. Think of the proteins in eggs; as they are heated, so they coagulate to form large aggregates that eventually become solid. If these same aggregates are allowed to begin to form in more dilute systems they will act as thickeners. Nearly all protein based sauces use egg proteins as the main thickening agent. This emphasis on egg proteins probably stems from the ready availability of eggs. However, some meat stocks contain significant amounts of denatured proteins (especially collagen) which can also act as thickeners although they coagulate in a different way and form gels as the liquid cools, rather than forming permanent networks as the liquid is heated. The recipe in Chapter 6 for oxtail soup relies largely on ensuring there are enough denatured proteins in the soup, which act to provide the thickening.

Basic Principles and Key Points

- When using egg proteins do not allow the sauces to overheat – too much crosslinking may occur and your sauce will become lumpy (rather closer to scrambled eggs than a fine sauce!). Monitor the temperature using a thermometer.

- Use a Bain Marie or double boiler to prepare the sauce where possible
- Always stir the sauce to prevent any overheating at the bottom or sides of the pan

There are very many different types of protein thickened sauces, to me the most interesting are the sweet sauces, such as custards, etc.

Egg Custard

Ingredients
4 egg yolks
500 ml milk
100 g sugar
Vanilla pod

Method

Separate 4 eggs and keep the yolks (you could use the whites to make a soufflé to serve with the custard – see Chapter 12 for details). Whisk the egg yolks and milk together in a bowl and stir in the sugar. Finely chop the vanilla pod and add to the egg, milk and sugar. Leave this mixture for an hour or so to allow the flavour of the vanilla to permeate through the mixture. When you are ready to cook the custard put a double boiler on to heat and, once the water in the bottom pan is boiling, turn down the heat and pour the custard mixture into the top pan. Make sure that once the custard mixture is being heated by the steam rising from the lower pan you keep on stirring all the time. If you have a suitable thermometer use it to monitor the temperature of the custard. Heat the custard until the temperature reaches 78°C and then remove from the heat and serve immediately. If you do not have a thermometer, you can assess when the custard is ready by observing how it thickens. If you lift your spoon out of the custard the mixture will run off quickly at first. As the custard starts to thicken, so it will "coat" the back of the spoon – that is to say a thin layer will remain on the spoon and not drain away. Once this happens the custard is ready to serve.

Recipe Variation

Ice Creams

Ingredients
4 egg yolks
500 ml milk
120 g sugar
Vanilla pod
300 g Fresh Fruit (strawberries, bananas, etc.) or other flavouring of your choice
500 ml cream (single or double as you wish)

Method

Make an egg custard as above and allow it to cool. Puree the fruit and add with the cream to the custard. Now freeze the mixture to make the ice cream. There are several methods you can use to freeze the mixture.

Freezing using a domestic freezer

The simplest method is to put the mixture in a suitable bowl and then put in a domestic freezer. You will need to take the mixture out every 10 to 20 minutes and stir it well, to make sure that the temperature is uniform and to break up any large ice crystals that may form.

Freezing using "freezing mixture"

A faster method that does not require a freezer is to use "freezing mixture" to cool the ice cream. Freezing mixture is made by crushing ice (about 1 kg is best) and adding to it about 200 g of salt dissolved in about 300 ml of water. Make this mixture in a plastic washing up bowl and stir it well with a wooden spoon. The temperature will drop rapidly to about $-12\,°C$ easily cold enough to freeze the ice cream. Now put the ice cream mixture in a metal baking tin and float that tin on the freezing mixture in the washing up bowl – take care that none of the salty freezing mixture accidentally gets in to the ice cream mixture. Stir the ice cream mixture regularly (every couple of minutes) and take care to scrape around the bottom and sides of the tin where the ice cream will start to freeze first. This scraping will prevent any large ice crystals from growing in the ice cream and ensure a smooth texture. The ice cream should be ready in about 20 or 30 minutes.

Freezing using an "ice cream machine"

There are two types of "ice cream" machines available. One comes with its own built in freezer. The other uses a double walled basin with some liquid in the sealed space between the inner and outer walls. This liquid is frozen by storing the basin in a freezer overnight. The melting of the frozen, trapped, liquid is used to extract the heat from the freezing ice cream mixture in the inner bowl.

The operation of either kind of machine is similar. The ice cream mix is placed in the bowl and a paddle (operated by an electric motor) rotates inside the bowl scraping the mixture away from the sides and thus preventing any large crystals from forming. Generally these machines can prepare about 1 litre of ice cream in about 20 minutes.

Making ice cream using liquid nitrogen as the coolant
The quickest method to make ice cream, and one I use for dinner parties, relies on being able to obtain liquid nitrogen. Liquid nitrogen is produced on an industrial scale and you will often see tankers carrying it around the country. Liquid nitrogen has many uses. For example, all scientific laboratories, as well as many food industries, use large amounts of liquid nitrogen for a variety of purposes; to run equipment that uses superconducting magnets, such as whole body scanners; to help achieve high vacuums; as a coolant; or simply as a source of pure nitrogen for use as an inert atmosphere in the packaging of perishable foods. So liquid nitrogen is readily available to industry. However, it needs special vessels to store it (and of course it will slowly evaporate even in the best containers) and the manufacturers will tend only to deliver it in large (100 litre or more) quantities.

However, if you know of a convenient source, and have a suitable vacuum flask (an ordinary Thermos is quite adequate) then you can make 'liquid nitrogen ice cream' for yourself. The method is very simple. First prepare an ice cream mixture of your choice. Put this mixture in a large metal bowl (don't use a plastic or glass bowl as there is a danger that the thermal shock of adding the very cold liquid nitrogen may fracture the bowl). The bowl should be sufficiently large that the ice cream mixture takes up less than a quarter of its volume (if you use too small a bowl the mixture will overflow and make an enormous mess when you add the liquid nitrogen). When you are ready to eat the ice cream put the bowl on a suitable table remembering to put some form of insulation, such as a ceramic tile or a cork mat, between the bowl and the table so that the table is not damaged as the bowl gets cold.

Before handling the liquid nitrogen make sure your eyes are protected in case any should splash into them – you really should wear proper safety spectacles. Next pour in the liquid nitrogen (the amount of liquid nitrogen should be about one quarter the volume of the ice cream mixture). Clouds of white 'fog' will rise up from the bowl as the nitrogen boils. You should gently stir the mixture to ensure the nitrogen is well distributed and that all the mixture becomes frozen, make sure you do this stirring with a long handled spoon that is well insulated – a wooden spoon is best. The time it takes to freeze the mixture depends on how much ice cream you are making. If you start with about 1 litre of mixture it should freeze within about 30 seconds. Larger volumes may take longer. If the mixture has not frozen fully when all the liquid nitrogen has boiled away, you should add a little more liquid nitrogen and repeat the stirring. Also, if all the ice cream is not eaten before it melts, you can simply re-freeze it using a little more liquid nitrogen.

Vinaigrettes and mayonnaise

Vinaigrettes and mayonnaise are basically sauces where very small oil droplets are kept in suspension in an aqueous medium. Such systems are conventionally called emulsions or colloids. There is a major branch of Physical Chemistry devoted to the study of the formation and stability of colloids. Indeed a great deal of the effort of this work over the years has been devoted to improving and understanding food colloids in general, and vinaigrettes and mayonnaise, in particular. Despite all these efforts over many years, and the major advances in understanding of the general principles involved, the making of a perfect mayonnaise in the kitchen still remains an art, rather than a science.

The reason for the difficulty lies more in the consistency of the ingredients, than anything else. In an industrial setting, the exact acidity and viscosity of the components such as the eggs, etc. can be monitored and controlled. However, at home we cannot easily make these measurements and have to rely on experience to ensure good results.

Here, I shall only give two general recipes for a basic mayonnaise and a simple vinaigrette. You will need to use a good deal of trial and error to perfect the use of these recipes, as their success or failure relies critically on the temperature and acidity of the ingredients.

In these recipes, the stability of the colloids is provided by 'soap like' molecules that sit at the surfaces of the oil droplets and keep them from separating – see panel for a more detailed description. In the mayonnaise these molecules are provided by the egg yolk and consist of some 'lipids' as well as some denatured proteins – see Chapter 2 for an explanation of how these molecules can stabilise colloids. In the vinaigrette, the stabilisation is provided by the ground mustard – the molecules here are much less effective, so the resulting mixture is significantly less stable and may need frequent remixing, if not used immediately.

Colloids and Emulsions

Colloids are mixtures of two liquids that are not compatible. One of the liquids forms small droplets (the discrete phase) that are surrounded by the other (the continuous phase). The liquid of which there is less, generally forms the droplets. In most colloids the liquids are oils and water. If you simply mixed some water and some oil together they will quickly separate into two layers with the oil floating on top of the water. In a colloid, the droplets need to remain stable for long periods. To produce a stable colloid a 'surfactant' is added. The surfactant is a molecule that has one end of which likes to be in the oil and the other which likes to be in the water. In an oil in water colloid, these molecules sit at the interfaces between the oil droplets and the surrounding water. The water 'sees' only the ends that like water and does not 'see' any of the oil; similarly, the oil 'sees' the ends that are like oil and hence does not 'see' any water. The simplest of such surfactant molecules are the soaps we use to wash away grease.

In foods, the surfactant molecules are often proteins, or parts of proteins, and can themselves be destroyed, or have their activity severely changed by small changes in the temperature and acidity of the surrounding water.

The stability of a colloid is very delicate, many small disturbances can easily upset the balance and let the oil drops 'see' the surrounding water. Once the oil sees the water, it will quickly start to separate out, at first into larger drops and then into a separate layer on top of the water.

Simple Mayonnaise

Ingredients
One egg yolk
50 ml spirit, or other fine vinegar
50 ml dry white wine
80 ml olive, or other good quality, oil
a little salt and pepper to taste

Method

Mix together all the ingredients except the oil in a bowl (or in a blender if available). Slowly drip in the oil while beating with a whisk as vigorously as you can (if you use a blender, simply add the oil slowly through the top). Within a few minutes the mixture should become quite stiff and almost white, it is then ready for use. If you need to store the mayonnaise, you will need to store it in the fridge to prevent any bacteria growing in the rich medium. However, cooling the mayonnaise can cause it to separate. If this happens you should be able to restore it by another vigorous beating once it has warmed up again.

Simple Vinaigrette

This recipe uses lemon and lime juice, rather than vinegar, to provide the sharpness in a salad dressing. I cannot use any vinegar at home since my partner has an allergy to acetic acid, so I have found many interesting substitutes over the years. The flavour of this dressing is a little unusual, but it provides a welcome, fresh edge to any salad. Of course, if you wish you can substitute any vinegar (but preferably a fine spirit vinegar) for the lemon and lime juices to obtain a more conventional dressing.

Ingredients
Juice of one lemon and one lime (about 100ml in total)
60 ml oil – I usually use a toasted sesame oil, but any high quality oil will suffice
$^1/_4$ teaspoon ground mustard seeds
$^1/_4$ teaspoon ground roast cumin seeds
ground pepper to taste

Method

Add all the spices to the juice in a bowl and then slowly add the oil while whisking vigorously. Keep on beating until the mixture thickens and becomes cloudy. The dressing is now ready for immediate use on any salad.

Fondues

Fondues are really just simple sauces made from cheese melted with some wine and flavourings, which are kept hot and used for dipping bread or other foods that become coated with the cheese sauce and eaten. There is, however, quite a lot of mystery concerned with the process of making fondues and a lot of folklore with the proper ways to eat a fondue. Of course, the only real rule should be if you like it then its just fine.

A fondue, like a mayonnaise, is a form of colloid. However, fondues are somewhat more complex as they have three separate parts, or phases, rather than the two components, or phases, in a mayonnaise. As we have already seen mayonnaise consists of two phases. The oil, or fatty phase consists of a large number of very small droplets of oil which are suspended in the aqueous phase which consists of water, with some acids and other flavourings dissolved in it. The whole colloid, or emulsion, is stabilised by special 'surfactant' molecules at the interfaces of the oil droplets and the water.

Fondues have a similar structure, except that there are two phases of droplets dispersed in the aqueous phase. It is the presence of this third phase and its tendency to separate into a gooey mess at the bottom of the pan that causes all the difficulties that sometimes arise when making fondues.

Both the fatty phase and the extra droplet phase come from the cheese used to make the fondue. As the cheese is melted so the fats in the cheese are released and form into droplets in the fondue. In most cheeses there are sufficient surfactant molecules present (left over from the milk from which the cheese was originally made) to stabilise these molten fat droplets and keep them in suspension in the fondue with little difficulty – all that is needed is a little stirring.

The third phase comes from the proteins in the cheese. As cheese is made so the (casein) proteins form a complex structure where proteins and calcium interact to form a porous sponge-like medium in which the fat resides. When the cheese is heated in the fondue pan, the fat melts and flows out of this protein complex, leaving behind aggregates of proteins and calcium. These aggregates then need to be dispersed in the fondue to form the second droplet phase.

If the proteins in the cheese have been well broken down (as the cheese matured) then they may be quite soluble and little problem will arise. However, in most cheeses the protein complexes will be quite insoluble and are likely to form into large aggregates that will sink to the bottom of the fondue where they will burn and form an unpleasant mess.

So the key to making a good fondue is to make sure that the protein complexes break up into smaller aggregates which remain suspended in the fondue and do not separate out and sink to the bottom. The problem in getting perfect results every time lies mainly in the variability of the ingredients. No two cheeses are alike and the degree to which the proteins are coagulated, or may have broken down will vary both with the type and with the age of the cheese. The acidity of the wine you use in the fondue will have a significant effect on the degradation of the proteins. The more acid the wine the more it will encourage the proteins to break up. Further, the presence of small amounts of citrates (e.g. from lemon juice) can greatly alter the way in which the proteins coagulate.

There are several useful tricks that you can use to aid in preparing a perfect fondue every time. You can add some acidity in the form of citric acid (from lemons, or preferably from Sodium citrate – available from health food shops) to help break down the protein complexes and stabilise them. If you have a choice of wines to use then chose the driest (i.e. the most acidic) which will also assist in increasing the overall acidity. Another remarkably effective trick, although one which is frowned upon by purists, is to add a little cornflour with the wine. The cornflour thickens the aqueous phase (increases its viscosity) to such an extent that the separation of the casein protein complexes is greatly reduced. At the same time as thickening the aqueous phase it seems that some starch can actually have an additional stabilising effect by helping to emulsify the fat.

Structure of Cheese

Cheese itself has a fairly complex structure which has its basis in the structure of milk. Some of the milk proteins (there are four particular proteins, collectively termed casein) form small aggregates called micelles. Each micelle is about 1 ten thousandth of a millimetre across. In the cheese making process these micelles become bonded together to form much larger aggregates which give cheese its firm texture. The proteins in these aggregates are bound together through strong mutual interactions with calcium and phosphorous ions. In milk the outside of the micelles is covered largely by one of the proteins (κ-casein) which is hydrophilic (water loving) and thus helps to stabilise the micelles in the surrounding water. In the cheese making process the hydrophilic parts of these stabilising κ-casein molecules are stripped away by the enzyme, rennin, causing the micelles to become unstable and form aggregates which link together in a loose structure throughout the cheese.

The casein protein aggregates form gels which can accommodate large amounts of water. These gels are themselves quite porous structures and are able to hold the fat particles from the milk as well as the water. The more the proteins in the gel phase are crosslinked to one another through the interactions with calcium ions, the firmer the cheese becomes.

Key Points when making fondues

- Always use at least one garlic clove in the fondue (the sulphur in the garlic can help to break up the protein aggregates)
- Use the driest available wine
- Add a little cornflour to ensure the protein in the cheese does not separate
- If available add a little Sodium Citrate (about 2% by weight of the cheese)

Basic Fondue Recipe

Ingredients

400 g	Cheeses (preferably use a variety of cheeses – a mixture of gruyere, cheddar and parmesan gives a good finish)
180 ml	Dry white wine
1 or 2	Garlic Cloves
20 ml	Kirsch
8 g	Sodium Citrate (if available), otherwise use lemon juice
10 g	Cornflour

Method

Grate the cheese, add the cornflour to about 20 ml of the wine. Cut the clove of garlic in half and wipe it around the fondue pan, crush the rest of the cloves and add to the grated cheese. Put all the ingredients in the pan and heat gently, stirring all the time. Once all the cheese has melted continue heating and stirring for a couple of minutes to allow the fondue to thicken. Serve immediately with a dry white wine. Use pieces of bread, sausage, celery, etc. to dip in to the fondue.

Some experiments to try in the kitchen

One of the most important aspects of making successful sauces is to be able to stabilise emulsions (or to make colloids); so here are a few demonstrations of emulsions you can try out for yourself to see what is possible. Of course some of the ingredients used here are not edible, so do not try to eat the products, just use them to give an idea of the sort of textures that are possible.

Oil – Water emulsion

Take a bottle with a screw top and fill it with about equal amounts of water and oil (any oil will do for this experiment – cheap cooking oil, or car engine oil – it doesn't matter: you won't be eating the results!).

Shake the bottle vigorously and then put it down on a table. Note that the oil forms into small droplets on shaking and that these droplets quickly coalesce and rise to the top when the bottle is left standing on the table. Note how long it takes for the mixture to separate into a single layer of oil on top of a single layer of water.

Now open the bottle and add a very little washing up liquid. Shake the bottle again and leave it to stand. This time the droplets should be much more stable and it should take a long time to separate – if you are lucky it may not separate fully. What you are seeing is the effect of the surfactant molecules in the washing up liquid. One end of these molecules likes to be in the oil while the other likes to be in water – so they act to stabilise the interface between the drops of oil and the water. However, the oil drops are still less dense than the water and most will slowly rise to the top. You may find that a few oil drops remain at the bottom and that a few small drops of water get trapped in the oil at the top. Nevertheless, you should find that over a period of an hour or so the mixture still separates into two distinct layers.

Now take the separated mixture, add a little more washing up liquid, and put it in an electric blender and whisk it as hard as possible for several minutes. The mixture should become quite white and thicken remarkably. The reason is that the fierce action of the blender has managed to break the oil into very small drops (the white colour indicates the drops have a size comparable to the wave-length of light – around one thousandth of a millimetre). Now if you leave this mixture to stand in the bottle it should take much longer to separate, if it ever does.

In the fine dispersion created in the blender, the oil drops are sufficiently small that they have "Brownian motion" – a random movement caused by the action of molecules of water colliding with the droplets. This Brownian motion helps to prevent the droplets rising quickly to the top. Also the increase in the viscosity of the emulsion – caused by the interaction of adjacent small oil drops – slows down any separation.

Once you have seen the wide range of thicknesses and stabilities that can be achieved with a 50:50 mixture of oil and water you might like to try some different mixtures (say 75% oil and 25% water) and see what happens.

Finally, you could try to apply your experiences from the above to making sauces, but you will need to find an alternative surfactant to the washing up liquid used in these experiments. Suitable surfactants occur in many foodstuffs e.g. egg yolks, (all cells contain amounts of lipids which are nature's surfactant molecules). In general seeds contain more surfactant than most other foods, for example ground mustard provides plenty of surfactant and can be used to stabilise a range of sauces. But do experiment for yourself – you should quickly find a range of workable thickening and stabilising agents for all your own sauces.

Sponge Cakes

Basic principles

Every cook has made many sponge cakes; I know I have. Among the cakes I've made over the years, there have been many memorable disasters; cakes that failed to rise, burnt cakes, cakes that collapsed when taken out of the oven, tough cakes and many others too numerous to mention. However, these days, now that I have learnt what is involved in making a good sponge, disasters are few and far between. In this chapter I hope to show you how, by understanding the basic scientific principles involved, it is truly easy to make a perfect sponge every time (or almost every time!).

A good sponge cake is moist and light; it 'melts in the mouth'. To be light the cake needs to be mostly air; it is made from a lot of bubbles. The 'melt in the mouth' character comes from the thin bubble walls dissolving rapidly in the mouth. In addition to lightness and a fine texture the sponge needs to be strong enough to bear the weight of a filling, such as cream and fruit, without collapsing. So what should the bubble walls be made from? You need something that has strength, but is still readily soluble in the mouth. To prepare a good sponge, you will need to make a foam of many small bubbles, in which the walls between the bubbles are strengthened with flour, and to add to the mixture flavourings and anti-staling agents.

The important questions are how to get lots of bubbles into the sponge? and how to keep them small? There are three traditional methods. The first and oldest, uses yeast, which is a micro-organism that lives by converting sugar into alcohol and generates the gas carbon dioxide as a by product. When using yeast the bubble size is controlled by kneading a stiff dough (how this works is another story related to bread making and is described in Chapter 8). But it is difficult to produce really small bubbles, and the yeast itself introduces flavours that are not always desirable. So this method for making the bubbles in sponge cakes is now only rarely used and is not generally recommended.

The second method is to use a raising agent, baking powder, added to the flour (or to use self-raising flour which comes with the baking powder already added). Baking powder is made from several chemicals which when heated in the presence of some water react together and produce carbon dioxide gas to form bubbles. As with yeast based sponges, the main problem is to keep the bubbles small. The solution is to make a stiff paste with butter, flour, eggs, and some water. When the mixture is put in the oven the rate at which carbon dioxide is produced needs to be slow enough, and sufficiently uniform throughout the mixture, that the mixture cooks while the bubbles are still small (that is before they start to burst). Then as more carbon dioxide is generated the bubbles expand until the supply of gas is exhausted, or the mixture is so stiff that the bubbles become "set" and any additional gas escapes to the surface. It is difficult to control both the rate at which the mixture cooks and the rate at which the carbon dioxide is released. It takes lots of practice to get consistently good results.

The third technique is to make the bubbles first, then add the flour to stiffen the bubble walls and finally cook the cake. The bubbles are made by beating eggs (either separated, or whole with added sugar) until they form a stable foam (or mousse). (The reasons why beaten eggs form stable foams is another story – see Chapter 12). Once the eggs are beaten into a firm stable foam the flour is folded into the mixture to give it some strength or 'body', and some butter added before pouring into the baking tin and putting in the oven. In this method you can control the size of the bubbles in the beating stage and the very small bubble size can be preserved throughout cooking.

In this chapter I describe in detail how the second two techniques (cakes made with raising agents and cakes made using egg foams) work, and suggest a few experiments and recipes for you to try for yourself.

Basic sponge recipe using baking powder (or self raising flour)

The most frequently used method to generate the bubbles in sponge cakes involves a 'raising agent' which generates gases by chemical means during the cooking. Some grades of flour (self-raising flours) come with suitable raising agents already added, while others (plain flour, general purpose flour) do not. The basic cake mixture consists of flour (plus raising agent), sugar, eggs and some fats. The ingredients are first blended together to form a stiff paste (or batter).

The cooking process is very complex, with many changes in the texture of the cake mixture and in the rate of generation of carbon dioxide gas occurring simultaneously. The baking powder starts to generate some carbon dioxide as soon as it becomes wet. However, the rate at which carbon dioxide is generated increases dramatically once the temperature rises above about 50 °C. When the eggs and other liquid are added to the flour, sugar and fat, some carbon dioxide is generated by the baking powder. This carbon dioxide starts to form very small bubbles within the mixture. Provided these bubbles are small enough and the mixture is stiff enough they remain stable.

When the cake is put in the oven much more carbon dioxide is formed and the small bubbles that were formed in the mixing process grow much bigger and some new bubbles are formed. To begin with as the temperature rises, so the cake mixture becomes runnier and less stiff. During this stage the bubbles can expand easily and will begin to rise to the surface. If the bubbles become too big, or the mixture is too runny, then the bubbles may burst through the surface and the carbon dioxide will be lost to the atmosphere. If too much carbon dioxide is lost the cake will not rise properly. Later, as the temperature gets above about 60 °C, egg proteins start crosslinking (see Chapter 2 for details) and make the mixture much stiffer. As the cake mixture stiffens so the bubbles become stable again. Further production of carbon dioxide now inflates these bubbles a little more, but, like rubber balloons, there is a limit to the amount of gas they can be filled with before they will burst.

The degree to which a cake will rise thus relies on the delicate interplay between several competing processes. The rate at which the baking powder can generate carbon dioxide over a wide temperature range, the total amount of carbon dioxide that the baking powder can produce, the rate at which the cake mixture heats up in the oven and the rate at which the egg proteins crosslink and stiffen the mixture, all affect the rising of the cake. With so many variables it is not possible to provide any detailed instructions to ensure success every time. Indeed, problems of lesser complexity are sufficient to leave an expert Chemical Engineer quite baffled. So it remains a mystery to me how so many cooks find making a cake by this method so straightforward!

Key Points to bear in mind when making cakes with self raising flour (or with baking powder)

- Always make sure the oven is preheated to the cooking temperature
- Make sure your baking tin is lined with butter and flour to prevent the cake sticking to the sides and base
- Do not leave the uncooked mixture standing for any length of time

Reasons why these key points are important

Once you have found a successful recipe it is always a good idea to make sure you follow it as closely as possible every time you make another cake. One of the most obvious things that will affect the rising of a sponge cake made using baking powder as a raising agent, is the oven temperature and the rate at which the cake gets hot enough for the egg proteins to begin to coagulate. It should be apparent that the sooner the egg proteins begin to coagulate the less likelihood there is of the cake collapsing through loss of carbon dioxide. So it is always best to put the cake in a hot oven – if the cake is put in a cold oven it takes longer before the proteins begin to coagulate with the risk that the cake could collapse. Similarly, since the carbon dioxide starts to be formed as soon as the liquid is added to the cake mixture, it is best not to leave the mixture standing around with the possibility of all the small bubbles that form at this initial stage growing larger and rising to the top of the mixture where the carbon dioxide may escape. Greasing and lining the cake tin is always a wise precaution against the possibility of the cake sticking and not coming out of the tin cleanly.

Ingredients

100 g	Flour
5 ml	Baking Powder (only if plain flour used)
100 g	Unsalted butter or other fats (80 g if using a food processor)
2	Eggs
100 g	Caster Sugar
up to 15 ml	Additional liquid (milk or water)

Method

Preheat the oven to 180 °C. Mix the sugar, flour (and baking powder if used) and butter together with a spoon or knife; keep working the mixture (either by hand or in a food processor) until the texture is one of crumbs about 1 – 3 mm across. Add the beaten eggs slowly beating the mixture all the time. As you add the eggs so the mixture should become quite stiff and semi elastic (that is if you stretch it then let go, it should not remain as it was but should tend to retract a little. Once the mixture reaches this stage stop adding any more liquid and keep beating the mixture for a minute or so. Finally, add more liquid as needed to obtain the 'right' consistency. It is impossible to put into words just what the required consistency is for this sort of cake. Indeed, the best consistency is different for different flours and when using eggs of different ages; for example when using strong, or high protein, flours a runnier texture is required. If the mixture is too stiff then the cake will not rise much; on the other hand if it is too runny then although it will rise in the initial stages of cooking it will collapse before it is fully cooked. To try to give some idea I would note that the mixture should remain stuck to a spoon when inverted (but only just); later in this chapter I suggest some simple experiments you can do to see for yourself what is the best consistency when you use your own ingredients.

Once the cake mixture is just right put it into a greased and lined tin (the quantity given should be about right for a 20 cm diameter tin) and cook at 200 °C for about 20 minutes. Test the cake with a knife to see if it is ready. If not, cook for a few minutes more and test again. When the cake is ready take it out of the oven, drop from a height of about 30 cm on to a hard surface (to prevent any collapse) and leave to cool before turning out of the tin.

Preventing cakes collapsing after cooking

The bubbles in a freshly cooked sponge are closed so that no air can get in or out; you might think of the cake being made up of lots of tiny balloons all stuck together. As the cake cools so the steam inside the bubbles condenses (changes back to water). Imagine all the tiny bubbles starting to deflate and getting smaller and smaller. Of course the cake will start to collapse. The cake is stiffer (more cooked) around the edges and supported by the tin so it is not likely to collapse too much there. However, it will collapse in the middle unless you can change the structure so that air can come back into the bubbles to replace the condensed steam. Dropping the cake, from a height of about 30cm on to a hard surface, passes a shock wave through the bubble walls and allows some of them to break, converting the cake from a closed to an open cell structure. Now air is able to get into the broken bubbles and the cake will not collapse.

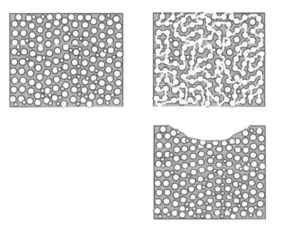

Figure 10.1. Top left, a freshly baked cake. Top right, the same cake after it has been dropped note the bubbles have burst allowing air in. Bottom right, if the bubbles are not burst the air inside them contracts on cooling and the cake collapses in the middle

A thought on dropped cakes
Once I gave a lecture demonstration on the science of cake making for my University's Department of Continuing Education. After the lecture the audience were all asked to fill in questionnaires so that the department could see how well (or poorly) I had given the talk and thus decide whether to ask me to give any more.
One question asked "what did you enjoy most about the talk?" One of the students gave the reply "When he dropped the cake". A few weeks later I had a lovely letter from the Head of the Department thanking me for the lecture and adding a note of sympathy for my apparent disaster in dropping the cake. Indeed he commented that it is always a pity that people enjoy it so much when a demonstration goes wrong.
To this day, he is still teased by those who know about his mistake in assuming I did not intend to drop the cake!

Why the recipe works

Purpose of the Ingredients

Baking Powder (already included in self raising flour)

The baking powder is the raising agent, without it the cake will not rise. Obviously it is an essential ingredient. Without baking powder the cake will not be able to rise. See panel for further details of how baking powder works.

Baking Powder
Baking powder is a mixture of chemicals, sodium bicarbonate and salts plus some starch to keep the mixture dry. Typically, two distinct salts are added, for example cream of tartar and Aluminium Sodium Sulphate. The salts become acidic when dissolved in water. The acids then react with the sodium bicarbonate to form a sodium salt and carbon dioxide. You can easily see these reactions in some simple demonstrations at home. If you put a little sodium bicarbonate (about a couple of teaspoons) in a bowl and add about the same amount of an acid (such as lemon juice or vinegar) you will see that the mixture fizzes as the chemical reactions produce carbon dioxide. You will also find that if you add plain water to baking powder a similar effect occurs – in this case there is no need to add acid as the cream of tartar forms tartaric acid as soon as it dissolves in the water.

The cocktail of chemicals in baking powder is chosen to ensure that the chemical reactions that produce carbon dioxide occur over a wide temperature range. Tartaric acid (from the cream of tartar) reacts with sodium bicarbonate at room temperature, but higher temperatures are needed before the reaction with Aluminium Sodium Sulphate sets in. Both reactions occur faster as the temperature is raised during the cooking process. Other acids present in the cake mixture will also react with the sodium bicarbonate at room temperature so that much of the sodium bicarbonate can be used up before the cake mix is even put in the oven. To compensate for this problem, some recipes that use acidic flavours such as lemon juice etc. call for additional sodium bicarbonate to be added so as to ensure a supply of carbon dioxide at the high temperatures during cooking.

Flour

The flour, in combination with the liquids present is used to make the batter or dough. The dough is beaten or worked to produce some gluten, a combination of two proteins found in wheat. It is the gluten that gives the batter its elastic properties which allow the gas generated by the raising agent to form bubbles that expand during cooking. The process can be likened to the blowing up of balloons; it is important that the batter has suitable elasticity so that it behaves like the rubber in the balloons and does not burst as it is inflated with the carbon dioxide generated from the baking powder.

Butter

The butter serves two purposes. First it acts as an anti-staling agent, for details of how and why fats delay staling see the panel on staling in Chapter 8. Secondly, and more importantly, the butter acts to keep the starch grains in the flour separate from each other so that when the liquid is added it does not form lumps. The process is directly analogous with the use of fats in the production of starch thickened sauces (see Chapter 9).

Eggs

The eggs provide the protein that coagulates during the cooking process to give the cake its final, firm, texture.

Sugar

In this recipe the sugar is mainly used as a flavouring agent. A cake with a perfectly good texture can be made without the use of any sugar; however, some adjustment of the amount of liquid added is necessary to compensate for changes in the runniness (viscosity) of the mixture. Further, an unsweetened cake would not be to most peoples' taste.

Purpose of the Instructions

Why mix the flour and fats together to make small 'crumbs'?

The starch granules in wheat flour contain a good deal of protein as well as starch. When any liquids are added the protein molecules at the surface of the granules absorb some water and the granules become very sticky. If the granules are not kept apart before the water is added then they often form lumps where the outer granules have stuck together and prevent any moisture reaching the inside. By mixing the flour with the fats the granules are coated with a thin layer of fat which slows down the rate at which they can absorb any water and prevent granules sticking together so avoiding the formation of any lumps in the mixture. If the fat is not 'rubbed in' then the mixture may not be very uniform and the final cake may have an unpleasant texture. Note that if a food processor is used to 'rub in' the fat then less fat is needed to achieve a good coating of the starch granules than working by hand. The food processor is more efficient at mixing. However, the food processor will also make a lot more gluten once the liquid is added so that a cake mixture made in a food processor may need a runnier texture than one prepared by hand.

Why beat the batter?

An important part of these cake batters is the 'gluten' that can be formed by stretching proteins in the starch granules of the flour. When two granules stick together the protein molecules at their surfaces become entangled and are stretched when the batter is beaten. The entangling and stretching processes cause the protein molecules to form 'gluten sheets'. These gluten sheets can behave rather like rubber. In particular they provide the batter with its elasticity and allow it to hold the bubbles as they are generated, rather than having them escape. By contrast, in cakes where the bubbles are produced by physical means (e.g. Genoese sponges) the formation of gluten is not necessary and is avoided by only folding in the flour gently at the end.

Why is it important to have the 'right' consistency of the mixture before putting it in the oven?

To achieve a uniform distribution of bubbles inside the cake it is necessary to ensure that new bubbles are formed inside the cake during the cooking process to replace those that are lost by the process of rising to the top. Baking powder has been developed over the years to provide a more or less continuous supply of carbon dioxide during cooking, so new bubbles are formed all the time. These bubbles expand as the internal pressure rises with the gluten sheets acting rather like rubber balloons.

In an ideal cake, at the early stages of cooking the viscosity and stiffness are so low that most of these bubbles rise to the top and escape. As cooking progresses the viscosity and stiffness increase and the bubbles remain trapped. Finally the mixture becomes so stiff that no new bubbles can be formed with the available carbon dioxide pressure. The key to producing a cake with a good texture is to start with a cake mixture of just the right consistency so that the new bubbles being generated throughout the cooking process balance those being lost at the surface.

In practice, the viscosity and stiffness of the mixture may too high, or too low, also there is a risk that all the baking powder can be used up well before the end of the cooking time. If the mixture is too runny, then the baking powder may well be exhausted before the mixture is stiff enough to trap any carbon dioxide bubbles, leading to a flat cake. On the other hand, if the mixture starts out too stiff, then there may never be enough carbon dioxide pressure to inflate any bubbles again leading to a flat cake.

Problem solving with sponge cakes made using a raising agent

Unfortunately there is no way to rescue most failed sponges once they have been cooked, so the guide below just shows you how to do better next time.

Problem	Cause and Explanation	Remedy
Cake does not rise	Insufficient or inactive baking powder.	Check the "best before" date of self raising flour and baking powder. Make sure you use the quantity of baking powder called for in the recipe.
	All the baking powder reacted before the cake started to cook.	The cake mixture may have been left standing, or it may have been too acid. Try not to leave the wet mixture standing before putting in a hot oven. If the mixture is acid (it tastes sour) then add a little extra baking powder.
	Mixture so sloppy that the bubbles of carbon dioxide are able to rise right through.	Far too much liquid used in the cake mix. Use much less next time.
Cake rises too much and over- flows the tin	Mixture too soft	Too much liquid used in the cake mix. Use less next time.
	Too much baking powder used	Use less next time
Cake collapses as soon as it is taken out of oven	Proteins not sufficiently coagulated i.e. cake not cooked long enough, or at high enough temperature.	Cook for a longer time and/or at a higher temperature.
Cake collapses in the middle	Caked cooked on outside but not in middle – cooking time too short, cooking tempera- ture too high.	Increase cooking time and reduce oven temperature.
Top well browned but inside under- cooked	Cake cooked at too high a temperature	Next time cook in a cooler oven.
Cake dries out or becomes stale very quickly	Cooking temperature too high, Cake cooked too long or no fat used in recipe.	Make sure there is some fat in the recipe. If necessary cook for a shorter time, or in a cooler oven.

Some differences you may find in recipes in cookbooks

Using Different Proportions

Many recipes use less butter and sugar than suggested above. The amount of sugar and fat may be reduced to half that quoted reflecting a range of cakes that can be made. Using more fat and sugar produces a richer cake. It is usual, but by no means essential, to keep the amount of fat and sugar equal. To find the cake that suits you best you should try out several variations on the basic recipe.

Using Different Fats

Many recipes use other cheaper (and perhaps healthier) fats to replace the butter. The cakes will all have slightly different flavours but it should make little other difference what fat you choose to use. However, if you decide to use some of the newer low fat spreads to replace the butter you should be aware that these spreads often contain little fat (as their names suggest) and a lot of water, so you will need to put in more of them than you might normally expect in order to get a reasonable texture in the finished cake.

Victoria Sandwich Method

In recipes for Victoria Sandwiches (and many other sponges) you are often told to begin by creaming together the butter and sugar, then add the eggs and finally add the flour. Such methods produce less gluten than the basic recipe above, since there is less opportunity to stretch out protein molecules joining together the flour granules. With less gluten, the mixture will be less elastic and there will be more chance for bubbles to escape during cooking. It is more critical to get just the right consistency for the mixture when using this technique. However, there are benefits; it is the gluten that makes for 'toughness' in the cake, so sponges successfully made by these recipes should have a better texture.

All in one methods

Some recipes suggest putting all the ingredients into a bowl together and using a power whisk to mix them. In such a method you rely on the power of the whisk to ensure that the flour does not form any lumps. This method certainly can work, but if you are at all timid in using the mixer on high power you could end up with some lumps in the final mixture. Further, if the mixture is too runny it is very difficult to make it stiffer, so when using all in one methods it is advisable not to add all the liquid until you can see how stiff the batter is going to be. I believe that although these methods can certainly save a little time, they are really only suitable for experienced cooks.

Extra ingredients – flavourings, salt, cream of tartar, extra baking powder etc.

A glance through any recipe book will provide many variants on the basic recipe for a sponge cake. In general there should be no need to add any extra ingredients except as flavourings; I include salt as a flavouring – if you haven't tried making a sponge without any salt you should do so; you may be pleasantly surprised!

In most cases there should be no need to add extra baking powder, cream of tartar (tartaric acid) or sodium bicarbonate (baking soda) beyond that already in the self raising flour (or the 5 ml added to 100 g of plain flour). However, if an acid flavouring such as lemon juice, or other fruit, is added then there may be some need of extra sodium bicarbonate (or in the absence of sodium bicarbonate, baking powder).

It is always a good idea to question whether all the ingredients suggested in a particular recipe are really necessary. If you know the main purpose of the ingredients you should be able to tell whether omitting something is likely to have an important effect. I believe that it is a really good idea to experiment a lot to find recipes that suit you best.

Some suggested variations etc.

Chocolate and other flavoured cakes

The possibilities for flavouring sponge cakes are limitless. The simplest involve adding some flavouring in with the flour. For example chocolate cakes are prepared by replacing a quarter of the flour with cocoa powder. Other flavourings that can be added, more sparingly, include cinnamon, or any other spices of your choice, a small quantity (about 2.5ml) should be added with the flour. Fruit flavours can be obtained by replacing some of the liquid with fruit juice or, if available, a few drops of essential oils.

Fruit cakes

The basic sponge recipe lends itself well to the addition of some dried fruits to make a light fruit cake. The fruit is simply stirred in just before putting the cake into the tin. When making a fruit cake – especially a cherry cake using glacé cherries – it is a good idea to use a rather stiffer mixture than usual to reduce the tendency for the fruit to sink during the cooking. A tip that can help if your cherries do sink is to put nearly all the cake mix in the tin and then mix the cherries with the remaining cake mix and just put them on top of the tin so that they fall through the cake as it cooks.

Steamed puddings

Steamed puddings, such as treacle pudding, can easily be made using the basic sponge recipe. The cooking is at a lower temperature (100 °C) for a longer time. Since the mixture will not get so hot it will not get so much runnier as it is heated. It is best to compensate by starting with a less stiff batter, either using an extra egg, or adding a little more milk. I suggest you use about 10ml more liquid ingredients than you use for a cake cooked in the oven.

To make a steamed treacle pudding, first grease a 500 ml mixing bowl; use a Pyrex or polypropylene bowl, not a polythene one as it might melt! (If in doubt pour some boiling water into the bowl; if the bowl becomes soft then immediately put it under the cold tap and run cold water over it for a few minutes; don't use it for steamed puddings). Put about 50 ml of golden syrup in the bottom of the bowl and then pour in the sponge mix. Put the lid on the bowl if you have one otherwise cover it with aluminium foil and put in a steamer; cook for 1 to $1^1/_2$ hours, turn out and serve hot.

Basic Genoese Sponge

Genoese sponge cakes do not use a raising agent such as baking powder, but rather the bubbles are made before the cake is put in the oven. The bubbles are in the form of an egg foam (traditionally made with whole eggs, but it can be made using only the whites). The other ingredients are then gently added to the stiff foam so as not to break any of the carefully prepared small bubbles and then the cake is cooked. The cake rises a little during the cooking as the air inside the bubbles expands and as some steam is generated.

This method of making a sponge cake is very much simpler to understand and to control, since all the bubbles in the finished cake are made, by beating eggs, before the cake is put in the oven. There are no complex chemical reactions producing any rising agents. The only reason a Genoese sponge will rise a little in the oven comes from the expansion of the air inside the bubbles and from the generation of a little steam as the cake heats.

The process begins with the preparation (by vigorous beating) of a very stiff foam of very small air bubbles in a egg and sugar mixture. Flour and melted butter are carefully added trying to avoid bursting any of these carefully produced bubbles. The mixture is then heated in the oven to crosslink and coagulate the egg proteins and so set the cake. Since there are few competing processes it is much easier to understand the process and to 'trouble shoot' any failures. In short this is a method that will produce perfect cakes every time.

Key points to consider when making Genoese Sponges

- Preheat the oven to the cooking temperature
- Grease and line the cake tin
- Beat the eggs and sugar to a very stiff foam
- Add the melted butter (or other fat or oil) last and put the cake immediately in the oven

Why are these key points important?

The best sponge cakes will be those with the most and smallest bubbles when cooked. Since all the bubbles in this type of cake are made by beating the eggs and sugar together, you should make sure you make very many very small bubbles at this stage. Egg foams are stiffer when they contain smaller bubbles, so you can judge the texture of your final cake from the stiffness of the foam as you beat the eggs – stiffer foams will produce finer textured cakes.

As you will see later in this chapter the addition of fats to this sort of cake poses a problem; the fats are added to reduce staling, but they also tend to burst the carefully made bubbles, so the quicker you can start the cake cooking after adding the fats the better. So again it is desirable to put the cake mixture into a hot oven as soon as it is ready.

As with any cake there is a risk that it may stick to the baking tin. Since an ideal sponge cake is very soft, especially while still hot, it is wise to grease and line the tin to guarantee it will come out easily and undamaged.

Recipe

Ingredients

100 g	Plain Flour
100 g	Sugar
4	Eggs
50 g	Unsalted Butter

Method

Preheat the oven to about 200 °C. Grease and line a 20 cm baking tin. Melt the butter in a saucepan. Beat the eggs and sugar with a power whisk until the volume has increased about six fold, and the mixture is stiff and very pale. Fold in the flour. Finally, quickly fold in the melted butter, immediately pour the mixture into the prepared baking tin and put into the preheated oven. After about 25 minutes test with a sharp knife to see whether the cake is cooked. If the cake is not ready continue to cook and test every 5 minutes. Once the cake is cooked, take it out of the oven and drop it (still in the tin) onto a hard surface from a height of about 30 cm. Leave the cake to cool for about 30 minutes and turn it out of the tin. Store the cake until you are ready to decorate it.

As the egg and sugar mixture is beaten so the volume increases, slowly at first but over about 10 minutes it will increase to up to six times its initial volume. Notice that at the beginning you can see some large bubbles, these get smaller as the mixture is beaten and eventually are too small to see. Also notice that during the beating, the colour of the mixture changes from deep yellow to a pale straw colour; in fact the mixture should end up almost white. The time all these changes take to occur depends on the temperature of the mixture, the size and the age of the eggs, the shape of the mixing bowl, the design and speed of the whisk and even on the way you hold the beater and bowl!

Problem solving with sponge cakes made using no raising agents

Unfortunately there is no way to rescue most failed sponges once they have been cooked, so the guide below just shows you how to do better next time.

Problem	Cause and Explanation	Remedy
Cake collapses in the oven	The bubbles in the foam collapse before the egg proteins coagulated during cooking. Possible causes are: the fat made the bubbles burst	The cake mixture may have been left to stand after the fat was added before it was put in the oven, or the oven may not have been fully heated before the cake was put in. Make sure the cake is put immediately into a hot oven as soon as the fat has been added to the mixture.
	The foam was not stiff enough in the first place	The mixture needs to be very stiff (rather like meringue mixture) before the flour and fats are added. Make sure you beat it long enough next time, or try beating over a bowl of hot water.
Cake does not rise	The mixture is so stiff that the bubbles cannot expand	The mixture that goes in the baking tin should be just about pourable. If too much flour is added it may become overly stiff – next time use less flour.
Cake rises too much and overflows the tin	Too much mixture was put in the tin	Use a larger tin or less mixture next time.
	The mixture was not stiff enough	Beat the eggs longer, or add a little more flour to stiffen the mixture next time.
Cake collapses as soon as it is taken out of oven	The egg proteins have not coagulated during cooking. – The cooking time is too short or the oven temperature too low	Cook the cake at a higher oven temperature for a longer time.
Cake collapses in the middle	The outside of the cake is cooked and the centre is not, because the oven was too hot.	Cook at a lower temperature next time.
	The cake was not dropped on a hard surface when taken out of the oven.	Make sure to drop the cake as described in the recipe when you take it out of the oven.
Top well browned but inside undercooked	The cake was cooked at too high a temperature.	Next time use a lower oven temperature.
Cake dries out or becomes stale very quickly	Too little fat was used in the mixture, or the fat was not mixed evenly.	Make sure you use enough fat and that it is well mixed.

How does the basic recipe work?

The purpose of the ingredients

Eggs

The bubbles are produced by beating the egg and sugar mixture to create a foam. The reason why beaten eggs can stabilise a foam is that the action of beating changes the shape of the egg proteins, a process termed denaturing. The denatured proteins behave rather like soap molecules and help to create foams (see Chapter 2 for details) just like bubble baths. In principle any source of protein could be used, but those in eggs are the most readily accessible and are particularly effective.

Sugar

Sugar increases the viscosity or 'thickness' of the whole eggs so that the speed of beating required to denature the proteins is achievable with a power whisk. Before power whisks were available it was necessary to heat the eggs at the same time as beating them, so making them easier to denature and so reduce the beating speed needed to denature proteins to that possible with a hand whisk. Without the sugar, it would be impossible to make a stiff enough foam, even with a power whisk, without heating.

Flour

The flour provides the starch which acts as a 'reinforcing agent' which stiffens and strengthens the egg foam. Without the starch the egg foams even when cooked would not be strong enough to support the cake filling, we would really have a soufflé rather than a sponge. There is also some cross linking of the proteins to the starch during cooking. Any fine starch powder would serve the purpose instead of flour, for example ground almonds can be used if a rich, white sponge base is required.

Butter

The butter acts as an anti-staling agent, (see Chapter 8 for details of how and why this occurs) it is not really necessary, but adds flavour and tends to provide for a richer texture. Certainly it is quite possible to make perfectly acceptable cakes without adding butter, or any other fats, but they won't last very long before becoming stale – typically a cake made without any fat will become stale in just an hour or two.

The purpose of the instructions

Why preheat the oven to about 200 °C and melt the butter first?

These preliminaries are best done before starting to make the foam. The beaten egg foam is fairly stable, it would be possible (but not advisable) to leave the beaten egg mixture for some time (maybe as long as half an hour) before adding the flour and most importantly the butter. Once the butter is added then the foam immediately starts to collapse – small bubbles coalesce to form larger ones. This process of collapse of the foam is directly analogous to the disappearance of the bubbles in the washing up bowl when you add greasy or oily dishes. Using melted butter ensures the time taken to mix in the butter is minimised, thus reducing the amount of collapse that occurs. It is very important to get the cake directly into the oven once the butter has been added. Therefore it is best to do all the preparation before starting to make the cake.

Why whisk the eggs and sugar until the volume has increased about six fold, and the mixture is stiff and very pale?

It is the beating of the eggs that provides the foam or bubbles that will make the cake light. This is the most important stage. It is difficult to describe in words just how stiff the beaten mixture should become, or to say just when to stop beating. In my experience the beating takes between 8 and 15 minutes. I have never found it possible to beat for too long. The best time to beat for is, I am afraid, one of those things which is so dependent on the actual equipment being used (and on the age and size of the eggs) that it is really a matter of experience.

Why fold in the flour?

The expression, 'folding in' is intended to convey the idea that you should add the flour gently so as not to break any of the bubbles you have just made so carefully. However, I have found that a little roughness is not too serious; for example I have made cakes where I have mixed in the flour with the electric whisk on its lowest power setting. I would not advise this procedure however until you are familiar with the whole process and are able to make good cakes every time. If the egg foam was not beaten enough, then this rough handling could lead to a significant increase in the size of the bubbles and hence to a poorer (coarser) texture in the final cake.

Why add the butter last and put immediately in the oven?

As noted above the addition of any fat, such as butter, causes the foam to start to collapse. It is therefore most critical that the cake should start to cook before this collapse becomes serious.

Some further Questions

Why does the mixture become pale?

You can use the colour of the egg sugar mixture during beating as a guide to the size of the bubbles, as the mixture becomes paler so the bubbles are becoming smaller. When we say something is blue we mean that it reflects back only the blue part of the light that falls on it, it absorbs all the other colours. When we say something is 'white' we mean that it reflects back all the light that falls on it. If there are particles whose size is similar to the size of a wave of light (about 1/2000 mm) then the light that falls on the particles will all be reflected back in all directions from the particles and the object will appear white. When there are more particles whose sizes are around a wavelength of light, objects will appear whiter.

In our cake mixture it is the bubble walls in the foam that have a size of about the wavelength of light, and they scatter the light back. The colour changes from yellow to white as we beat the eggs because the size of the bubbles is being reduced and more bubbles are being formed.

How stable is the mixture after it is beaten?

Until we add the flour the beaten mixture will be quite stable, we could leave it for maybe half an hour without too much collapse: the bubbles will coalesce so the bubble size may increase a little, but this is not a major difficulty. Once we add the flour the fine particles will puncture many of the bubbles so the bubble size is immediately increased. As soon as we add the fat the bubble interfaces are de-stabilised and the mixture starts to collapse. The only way to reduce this tendency to collapse is to cook the egg proteins as quickly as possible. Once the egg proteins start to cook they form chemical bonds that stabilise the foam permanently.

Some differences you may find in recipes in cookbooks

I have seen many variants on this basic recipe in cookery books. In most cases the differences are unimportant, often the recipes given are more complicated than really necessary and in a few, following the recipes could lead to a real disaster! I have selected a few common instructions given in recipe books to discuss in more detail.

Separate the eggs and beat only the whites

It can be difficult to whisk whole eggs to create a stable foam. Indeed, in all recipes for meringues etc. you will find the instruction to separate the eggs and to avoid leaving even a speck of egg yolk in the whites; because, it is said, even the tiniest amount of egg yolk will prevent the whites from whisking up to a stiff foam. The problem with whisking whole eggs is that there is some fat (cholesterol) present in the egg yolks. Any fat makes it harder to whip up a foam. However, as you will have already noticed, the majority of recipes for a Genoese sponge involve whipping whole eggs (with the addition of some sugar) to make a stiff foam. It requires a lot of energy to whisk up whole eggs as you have to denature a lot more proteins to make lots more soap like molecules to coat all the fat. To make it easier to create a stable foam, some recipes suggest you whisk the egg whites on their own and then fold in the other ingredients including the egg yolks. In such recipes it is important to add the yolks at the end, just before the melted butter since the fat in the yolks will tend to destabilise the foam just like the butter in the basic recipe. There can be two advantages in whipping the egg whites separately. First, if you do not have a power whisk it will greatly reduce the effort involved; and secondly, if you want to avoid adding sugar it is almost impossible to use whole eggs to make a foam.

Whip the eggs over hot water

Before the advent of power whisks it was common practice to heat the eggs over a pan of boiling water to assist in the denaturing of the proteins. I have used this method and always found that I have scalded myself with all the steam coming from the pan of boiling water. I do not recommend using this technique; if you do not have a power whisk the methods that use separated eggs are much easier and you are far less likely to injure yourself! Also, you must avoid overheating the eggs while whisking or all you will have is scrambled eggs!

Adding the fat during the beating of the eggs

I have seen several recipes that suggest adding the melted butter (or in some cases oil) to the eggs while beating them. I have tried this out and have only met with disasters; the foam collapses so quickly that I have found it impossible to make an acceptable sponge.

Using self raising flour

Some recipes call for self raising flour. If the bubbles are properly made, then there should be no need for the use of a raising agent. Indeed, the cake may rise far too much and the bubbles be so large that the texture becomes coarse and unacceptable.

Using a copper bowl

Many older recipes suggest you beat the eggs in a copper bowl. This can have two advantages that may make it easier to beat the eggs by hand. First, copper is a good thermal conductor, so if you hold the bowl in one arm while operating the whisk with the other, some of your body heat will be transferred through the bowl to the eggs raising the temperature and making it easier to denature the egg proteins. Secondly, copper ions can interact with proteins to link them together making it a little easier to form a network and thus a stiff foam. However, neither of these advantages are important if you use a power whisk.

Some Variations and finished cakes to try

Black Forest Gateau

For a Black Forest Gateau you will need a chocolate flavoured base so about one ounce of the flour is substituted with cocoa powder in the basic Genoese recipe; the cake is still prepared and cooked in the same way as above. To finish the Black Forest Gateau, you must first stone some cherries (or buy ready stoned cherries); you will need about 500g of cherries for the gateau (fresh cherries are best, but tinned ones are quite acceptable). All but 13 of the cherries should be cut in half. Leave the cherries to dry on a piece of kitchen paper.

Cut the cake base into three layers and mark them on the side with a dab of cream so that you remember their orientation when re-assembling the gateau. Take the first layer and sprinkle some kirsch on it, then spread a layer of whipped cream (use 1 pint of whipping cream for the whole gateau) over the base. Next put half the halved cherries into the cream and spread some more cream over these cherries. Put the middle section on the cake, making sure you get it in the correct orientation using the marks you made when cutting the base. Add kirsch, cream and cherries as before. Put the top on the gateau, spread some cream all over the top and sides.

Decorate the sides of the cake with grated chocolate or chocolate vermicelli. I have seen recipes that tell you to roll the cake in grated chocolate to cover the sides; this method can work, but more often than not the inexperienced will end up with the cake falling to pieces and becoming an almighty mess! A simpler solution is to fill a dessert spoon with the chocolate and run it up and down the sides of the cake while tipping the spoon slightly so that the grated chocolate falls onto the cake. The process leads to a lot of chocolate falling off. You can either collect and re-use this chocolate or just eat it! To finish off the Gateau pipe 12 rosettes of cream evenly around the top and one in the middle, put a whole cherry on top of each rosette. Eat and enjoy!

Strawberry Gateau

A Strawberry Gateau is made in much the same way as a Black Forest Gateau. The basic Genoese recipe is modified by exchanging about 25 g of the flour for about 50 g of ground almonds to make a flavoured base. You need to use more ground almonds than flour due to the coarser nature of the ground almonds. Once cooked and cool, the base is cut into two or three layers and then the cake is assembled in the same way as for the Black Forest Gateau, using fresh strawberries cut in half instead of cherries to fill between the layers and using Madeira rather than Kirsch to sprinkle on the layers before assembly. The sides of the cake are finished with chopped almonds and the top decorated with piped cream and whole strawberries.

Too Rich Chocolate Cake!

Every year on my partner's birthday, I have to make a cake for her colleagues to eat in their tea break. It seems that all people who work with computers adore chocolate! Over the years I have tried to make a cake that is so rich that they cannot finish it – unfortunately I have failed, but I have come up with several variations that are appreciated by true chocoholics! A favourite example is based on a Genoese sponge, with an extra pastry base.

The pastry base is made from a variant on short crust pastry. Make a 100 g quantity of shortcrust pastry using your usual recipe but replacing a quarter of the flour with cocoa powder; roll out this pastry to a thickness of about 3 mm. Cut out a circle the same size as your cake tin (use the tin as a guide!) and cook it on a baking sheet. If you are used to making sweet French pastry then you will get a richer and more crumbly effect by using that recipe. Next prepare a Genoese sponge replacing a quarter of the flour with cocoa powder. Cut the sponge into two layers.

Once the cake and pastry base are ready prepare some rich chocolate butter cream. For a 20 cm cake you will need 250 g icing sugar, 250 g unsalted butter (here it is advisable not to use salted butter as the salt makes for an after taste in the final product), 50 ml milk, and 150 g chocolate (use a high quality cooking chocolate with at least 45 % cocoa solids). Begin by creaming together the butter and sugar, if you have a food processor this is simple – just put the ingredients in the bowl and blast away until you have a smooth thick cream. If you have to do this by hand make sure the butter is already soft or you will get a lot of exercise! Next melt the chocolate and add the heated milk. Mix the chocolate into the butter cream. Try not to eat too much at this stage!

Now spread a generous layer of the butter cream over the pastry base and sprinkle as many chocolate chips as you can on to it. Press the chocolate chips down with a palette knife and spread a thin layer of butter cream over the top. Next put the first layer of the Genoese on top and spread over it a generous layer of butter cream. Then put on the top layer being careful to keep it in the correct orientation so as to finish up with a level topped cake (see instructions in the section on Black Forest Gateau to see how to do this). Now spread the remaining butter cream over the top and sides of the cake. Cover the sides with chocolate biscuits of your choice. The top can be finished off either with a solid piece of chocolate (made by spreading melted chocolate on sheet of aluminium foil and cutting to size using the cake tin as a guide when almost set) or with miniature chocolate confectionery bars of your choice.

Some general issues that arise when baking sponge cakes

Using salt and/or salted butter

Salt is not necessary for making a sponge; however, some people seem to find the flavour it imparts pleasant. I am not amongst these, in my opinion any saltiness detracts from the flavour of a cake. The use of salt is a personal decision; you should try cakes with and without salt and decide what you prefer. In all the recipes here I specify unsalted butter, in most cases this is just my personal preference; if there is a good reason (other than taste) not to use salted butter I say so.

Sifting flour
Nearly all recipes I have ever seen that call for the use of any flour tell you to sift the flour before you use it. The only reason given in modern cookery books is that it aerates the flour. However, usually when you use flour you add liquid and make a paste that excludes any air. In fact the real reason for sifting the flour is historical. Years ago, before the advent of supermarkets, grocers sold flour loose. It was possible for foreign bodies (weevils, the flour moth and its larvae and even the occasional mouse droppings etc.) to get into the flour in the bins at the grocery; so you always passed the flour through a sieve to avoid serving weevil or mouse dropping flavoured cakes! In general there is no longer any need to sift flour before use. The only times when it can be useful are when you want to be absolutely sure there are no 'lumps' in the flour. Lumps are rare these days as most domestic flour has additives to prevent the separate grains sticking together.

Testing to see when cakes are cooked

Recipe books recommend many different techniques to test when a cake is cooked. The methods mostly rely on the principle that all the egg proteins have coagulated (solidified). If there are still any proteins that have not coagulated then the mixture will be a little sticky, like the yolk of a soft boiled egg. However, once all the proteins have coagulated then the mixture will no longer be at all sticky, like the yolk of a hard boiled egg. Of course, it is best to avoid cooking a cake too long which will give a dry, even a burnt, texture. The test I use is to push a knife into the middle of the cake. If the cake is cooked the knife will come out clean, since the mixture won't be 'sticky'. To avoid any risk of overcooking it is best to start to test the cake before you expect it to be cooked and then keep on trying every few minutes until it is just ready. Some people may object to using this method since there is a risk of the knife leaving a mark. I find that any marks made by the knife in testing before the cake is cooked usually heal over as the cake cooks further and only one small mark is left by the final test. This mark should not matter since most sponge cakes are filled and decorated before serving, so that the decoration will disguise any mark you may have made.

Why are you told to leave the cake to cool before taking it out of the tin and to store it until you are ready to decorate it?

When a cake comes out of the oven it is still hot and soft. While it remains hot it continues to cook and become a little stiffer. As it cools, it becomes even stiffer and more able to stand its own weight. Although it should be possible to turn the cake out immediately a short period of cooling will ensure there are no problems. The texture of all baked goods continues to change as they are stored (see the comments about staling and the addition of butter earlier). Many cooks find it best to store the cake base for a few hours to let it become firmer before decorating.

Experiments to try for yourself

Experiments to determine the 'right' consistency for the cake mixture

The best way to learn the right consistency for your ingredients is by trial and error. I suggest you perform a simple experiment. Try making a set of small cakes in a bun tin, using a mixture with a different consistency in each space in the tin. First, make up a basic mixture using 50 g self raising flour, 50 g fat, 50 g castor sugar and $1/2$ egg. Put 25 g of this mixture in the first space in the tin. In another bowl beat together the rest of the egg with 20 ml water or milk. Next beat 5 ml of the liquid into the remaining cake mixture and put 25 g into the next space in the bun tin. Now add another 5ml of liquid to the remaining cake mix and put another 25 g of the mixture into the next space. Keep repeating until you have used all the mixture (this should give you 8 small buns). Try to remember the consistency of each mixture as you put it into the tin, this will help you judge the sort of mixture you want in the future. Cook in the usual way. After cooking test all the cakes (by eating them!) to see which one you like the best and then use a mixture of that consistency in the future. As a guide to help you if you use these experiments to improve your sponges, I have listed in the table the approximate amount of liquid that would be needed when using a mixture of 100 g flour for each of the buns. You might like to carry out more experiments using different quantities.

Bun Number	Number of eggs and amount of liquid (beaten eggs and milk) added to a mixture made with 100 g flour)
1	1 egg
2	1 egg + 7 ml liquid
3	1 egg + 14 ml
4	1 egg + 25 ml
5	2 eggs
6	2 eggs + 15 ml
7	2 eggs + 30 ml
8	3 eggs + 50 ml

An explosive experiment to demonstrate the generation of Carbon dioxide by baking powder

For this experiment you will need an old 35 mm film container – one of those little plastic boxes the films come in, some baking powder, and some water. If you can also get some bicarbonate of soda and some acid such as vinegar that would be a good idea.

Put about 5 g (half a teaspoon) of the baking powder in the plastic box and fill about $^1/_4$ full with water. Quickly snap on the lid and leave on a work-top. As the water mixes with the baking powder, so some carbon dioxide will be generated. As more and more carbon dioxide is formed so the pressure inside the plastic container rises, until it becomes sufficiently high to blow the top off in a small explosion. Try varying the amount of baking powder and water you put in the container and see how the time it takes for the explosion to happen changes. Also try adding hot, rather than cold water and see how that changes things. With a little practice you should be able to predict quite accurately how long it will be before the top is blown off, then it is a good time to impress you friends by timing a short explanation of what is happening ending with "… and then there will be a sudden bang!" just as the pot goes pop!

If you have some bicarbonate of soda (sodium bicarbonate) and acid you can get a stronger reaction by adding either or both to your baking powder. You can investigate how the acidity of the water added affects the reaction, etc.

An experiment to demonstrate that egg foams are collapsed by oils

First make a foam using some bubble bath – just whisk together some bubble bath and some water using an electric whisk – use about 5 parts water to one part bubble bath and you should be able to make a lot of foam. Now drop a little oil (or melted fat) on top of the foam. Notice how it destroys the foam as it quickly eats downward.

Now separate some eggs and make a foam from the egg whites and repeat the same experiment. Which collapses faster when oil is added, a bubble bath foam, or an egg white foam?

Pastry

Introduction

There are many different types of pastry; all of these pastries are based on a simple flour and water mixture usually with some added fat, and sometimes incorporating other ingredients such as eggs or sugar, to make a richer product. The ingredients are thus very similar to those in breads and cakes, except no raising agents are used in pastries. However, the final products are very different from breads and cakes. In cakes the use of raising agents together with the inclusion of eggs allows the starch particles in the flour to become "glued together" to form large aggregates, while at the same time retaining an overall light texture from the inclusion of the many small air bubbles. In breads, again the inclusion of small air bubbles contributes significantly to the difference in texture, and the use of yeasts as raising agents contributes to the distinctive flavour of the bread. Also, in bread making, one tends to maximise the formation of gluten during the kneading stages, while in pastry making one generally, but not always, tries to minimise the production of gluten.

There are essentially four types of pastry. The simplest, shortcrust pastry, is a crumbly product, used in many applications, but mostly as a covering for both sweet and savoury pies. Pie crust pastry is a strong and stiff pastry used for raised pies where the pastry itself forms the dish to hold the pie. Choux pastry is a light pastry where steam generated within the pastry as it cooks expands and opens up fairly large cavities inside the pastry. Finally, there are the layered flaky and puff pastries which have many uses, for example in sausage rolls and cream puffs. In most cases, to keep the pastry from being tough, it is important to keep the gluten content to a minimum. However, since it is usually necessary to knead the dough to form it into a suitable shape, some gluten formation is inevitable. The art of good pastry making relies on using techniques that keep the pastry dough in a suitable handleable form, while as far as possible preventing gluten formation.

Hydration and Gluten formation

Gluten is formed only when wheat flour is deformed in the presence of water. The water serves to partially dissolve, or swell, the proteins that are already present in the flour. The process of swelling these protein molecules with water is often referred to as 'hydration'. This hydration is just the first stage in gluten formation. Once the protein molecules have been hydrated, they have to be at least partially denatured by mechanical manipulation and then they are able to reform into new structures where several proteins come together. There are two distinct components in the gluten. Gliadin is a sticky, semi-fluid protein that is soluble in alcohol; glutenin is not soluble in alcohol and has a fibrous elastic texture. Gluten is formed when lots of links are created between many different gliadin and glutenin proteins. The resulting aggregates of proteins are often referred to as protein complexes.

Of course, before these protein complexes can be formed the proteins that are to come together need to be well hydrated; in fact about twice as much water by weight of the proteins is needed to form gluten. This requirement for a large amount of water to hydrate the proteins provides a key to the control of gluten formation. By controlling the rate at which water is added, and the extent to which the water can reach all the starch granules in the flour, the overall degree of hydration of the proteins can also be controlled. For example, rubbing in fats to coat the outside of individual grains of flour will tend to repel water from the surface and thus decrease the rate at which hydration can be achieved. Hence the use of this technique in pastry making helps to control the amount of gluten formation.

Once hydrated, the different protein molecules can begin to interact with one another and new inter-protein bonds (di-sulphide bridges, hydrogen bonds, etc.) can be formed. The detailed way in which these processes occur is not well understood, but it is well known that some stretching of the individual proteins helps to break the initial internal bonds. As a flour and water dough is kneaded so the gluten progressively develops. The proteins between the starch granules develop into thin sheets which break into fine fibrils that stick to one another and stretch as the dough is kneaded. If the dough is allowed to relax before the gluten is fully developed, then some of these stretched proteins will relax back without forming the new inter molecular bonds that stabilise the gluten. This is why pastry recipes call for rest periods between rolling (or kneading) and tell you to avoid over working the dough.

Basic Principles common to all pastry making

Probably the most important consideration when making pastry is to control the elasticity of the dough. If the dough is too elastic, it will snap back to its original shape when rolled out. On the other hand, if the dough is too plastic it will flow easily and will not be able to hold its own shape after moulding. If you are not familiar with the terms "elasticity", "elastic" and "plastic" you will find a brief description in the panel below. The elasticity of the dough is determined in large measure by the moisture content and the level of gluten formation. Drier pastry doughs will generally be stiffer and a little more plastic, while those with a high gluten content will be very elastic. So, to be sure that you can keep the texture of the dough under control, you need to be able to keep the gluten content and the water content under control. It is fairly straightforward to control the water content – simply by measuring the amount of water added to the flour and fat mixture. If too little water has been added, the dough will be so stiff that it breaks up when you try to roll, or otherwise process it – you can simply add a little more water and carefully knead it into the dough. On the other hand, if there is too much water so that the dough sticks to the rolling pin, or flows under its own weight, etc. then there is again a simple remedy – just add extra flour to bring the dough back to the required consistency. Of course, it is only with practice and experience that you will learn just what this 'right' consistency is for any particular type of pastry.

Elasticity and Plasticity

Scientists use many words with which we are familiar but give them somewhat different meanings. Elastic, plastic, elasticity and plasticity are all examples of words in common usage, but where the actual scientific meaning can be rather different from the normal usage.

Most of us think of "elastic" as the stuff that keeps our underwear from falling down! However, to a scientist a material is "elastic" if after being deformed it returns to its original shape once the deforming force is removed. Thus under the scientific definition knicker elastic is indeed an elastic material. "Plastic" to most people is a word that describes a class of modern materials such as polyethylene that we use for a variety of purposes – mostly as packaging. However, the scientific meaning (which is much closer to the original meaning before the "plastics" were invented) is that a "plastic" material will not return to its original shape after being deformed. The deformation will be permanent. Thus, for example, plasticine and clays are "plastic" materials.

Most materials are actually both elastic and plastic depending on the amount of force used to deform them. If you have a small accident with your car and drive it slowly and gently into a wall the bumper will be deformed, but will recover its shape when you back away –the small deformation induced by this small force is elastic. However, if you should have a more serious accident and drive quickly into a wall, the damage to the bumper will be permanent. The larger deformation under the greater force is plastic.

So it is with pastry doughs – when deformed they will recover partially toward their original shape. The part of the deformation that is recovered is called the "elastic" deformation while that which is permanent is called "plastic".

A complication arises in that the recovery from deformation is not necessarily instantaneous. A dough that has been rolled out may slowly shrink back with the maximum recovery taking several minutes. This property of materials to recover slowly is known to scientists as "viscoelasticity". A viscoelastic material, if deformed very slowly, will flow so that the deformation is plastic, whilst the same material deformed quickly will usually be quite elastic.

The degree to which a viscoelastic material will show elastic or plastic deformation depends on the amount of deformation, the speed of the deformation and on the temperature. The recovery of deformation is much faster at high temperatures (and slower at low temperatures). So, if you do have that unfortunate accident with your car, it is worth remembering that most bumpers these days are made from a viscoelastic material (glass filled polypropylene) which will recover from deformation very much faster when heated. You may well find that an apparently permanently dented bumper that you are ready to replace can be restored simply by warming it with a hair drier.

Managing to keep the gluten content low is more difficult. First, you need to control the extent to which the starch granules absorb water. Secondly, you need to limit the amount of stretching that the hydrated granules are subjected to. Remember that once the gluten complex has formed it is more or less impossible to get rid of the gluten, so the most sensible thing is to minimise gluten formation in the first place. If and when you do want any gluten to form you can do so at the required time.

The first stage of gluten formation is the "hydration" of the proteins surrounding the starch granules. If you can reduce the amount of water reaching and swelling these proteins, you will reduce the potential for gluten formation later when handling the dough. There are two important considerations. First, you can rub the fat into the flour before adding any water. The fat coats and even to some extent penetrates the starch granules making it difficult for water to get to them. Obviously, the more you rub the fat into the flour and the finer the texture of the resulting mixture, the better the dispersion of the fat around the starch granules will be, so it will be more effective in reducing gluten formation. Secondly, you can try to reduce the degree to which the water penetrates into the starch granules and swells up the surface proteins. The solubility of the proteins is rather sensitive to the temperature, at lower temperatures the proteins will absorb much less water. So keeping the flour and the water you use as cold as possible will reduce the hydration of the starch granules and will thus assist you in controlling the gluten content.

Finally, once the pastry has been made and rolled out it is a good idea to allow it to stand for some time (typically 10 to 20 minutes) before finally cutting it to shape and using it. The stress built up by the rolling action will slowly relax and the rolled sheet will slowly contract – the more elastic the dough the greater the extent of this contraction will be and the faster it will occur. If you do not allow the rolled dough to stand then it will contract during the cooking and will change its shape then. So, for example, if you have carefully made a circular flan case, but have not allowed the rolled out pastry to stand for some time before cutting the circles out, it will end up after cooking as an elliptical case! Similarly if you are using the pastry for a pie crust and forget to leave it to stand it will shrink in one direction on top of the pie and may break up perpendicular to that direction.

Commercial pastry

These days you can buy ready made pastry of all forms in every supermarket. So you may wonder whether it is really worth the effort of making your own pastry. Of course, if you are in a hurry, these ready made pastries are excellent. However, they are generally made with cheaper fats (or hydrogenated vegetable oils) and tend to have much less flavour of their own than home made pastry made with butter.

These commercial pastries do have other advantages, they are very consistent and can tolerate a wider temperature range than most home made pastries. Commercial puff pastry, in particular, can be much more uniform in the way it rises, than the home made version.

Key points to bear in mind when making pastry

- Always rub the fat into the flour before adding any liquid ingredients
- Make sure the fat is well rubbed in to the flour – the mixture should have a fine texture with the small pieces of flour bound together by the fat having a size of about 1 mm or less
- Keep the flour mixture cool and only add cold water (except when making raised pie pastry)
- After rolling or kneading the dough allow it to rest for ten to twenty minutes before using

Some reasons why these key points are important

Rubbing in the fat provides a hydrophobic layer around the starch granules and helps to prevent water from swelling the proteins surrounding them. Remember this hydration is the first stage of gluten formation, so it is important to keep the flour quite dry until all the fat has been rubbed in. Also remember that cold water is less prone to hydrate the starch proteins so keeping everything cool will also help to reduce the possibility of forming the tough gluten that will spoil the texture of the pastry. Allowing the pastry to rest allows any elastic deformation to recover so that the pastry will keep its shape when cooked, it will also allow any proteins that have become stretched, but not yet formed into gluten to relax back.

Short Crust Pastry

The objective is to make a fine, friable textured final product that crumbles in the mouth, so it is very important to prevent as far as possible any gluten being formed as the flour is rolled and kneaded. Since gluten forms when hydrated starch granules are stuck together and then stretched out, you must minimise the formation of the gluten by reducing the hydration of the starch granules and by trying not to stretch the dough too much.

So in addition to the key points above you need to be most careful to keep any stretching of the pastry dough to the bare minimum. Remember that the dough will be stretched at all stages including the initial mixing. So mix very gently using a tool with a thin edge (note, a metal spoon will be better than a blunt wooden spoon) and try to squeeze the dough together in your hands. Further stretching will occur when the dough is shaped, or rolled out, so be careful to use as little force as possible and leave the dough to rest and recover after each rolling. This period of rest and recovery allows any stretched proteins that have not already formed a gluten complex to relax before they get the chance to form gluten.

Basic Recipe

Ingredients
300 g Plain Flour
150 g fat (butter, or lard and butter mixed – do not use "low fat" spreads as they contain a lot of water and rubbing them in would promote gluten formation)
50 ml cold water
2 g salt (optional – salt will add a savoury flavour to the pastry so use sparingly and only for savoury pies, etc. There is no need – apart from flavour – to add any salt)

Method

Begin by making sure all the ingredients are cold (preferably taken straight from the fridge). Put the flour in a bowl, again it can be a good idea to cool the bowl; if you are using salt add it to the flour. Cut the fat into small pieces, add it to the flour and work it between your finger tips until it is well "rubbed-in" and there are no lumps of fat left. The fat will make the flour form into small 'crumbs'. You should aim to keep these crumbs as small as possible – about a millimetre is an ideal size.

Once the fat has been rubbed in to the flour you can proceed to add the water to form the pastry dough. Remember once you start to add water you will be hydrating the starch granules and any stretching of the hydrated granules will lead to gluten formation and a tough, rather than crumbly, pastry. It is important that the flour is evenly hydrated – if some parts have a much higher water content than others then water from the wetter parts may boil during cooking leading to blistering of the surface of your final pastry. So the best way to add the water is a little at a time just sprinkled over the surface of the flour and fat mixture in the bowl.

After each addition of water, gently stir the mixture and turn it so that drier material is at the surface. Try to avoid any squeezing or stretching motions during this stirring process; you may find it easiest to use your hands to turn over the mixture, or you may prefer to use a knife, or metal spoon with a reasonably thin edge, for the purpose. It is best to avoid using a blunt tool such as a wooden spoon as you will have less control and are more likely accidentally to stretch the dough.

As soon as the mixture begins to stick to itself stop adding any water (the exact amount of water needed will depend on many factors including the humidity in your kitchen, the ambient temperature, the type of wheat the flour was made from, etc. so care is needed not to add too much water). Next, using your hands gently draw all the mixture into a single ball and very carefully squeeze it to form a coherent dough. During this drawing together of the dough, you will inevitably form a little of the dreaded gluten – indeed a little gluten at this stage is a good thing as it helps to keep the pastry dough in a coherent ball.

Once the pastry dough has been prepared set it aside in a cool place for 15 to 20 minutes to allow the partially stretched proteins between starch granules to relax and so reduce the degree of subsequent gluten formation. When you are ready to use the pastry you will need to roll it out to the required shape and size. Since more gluten will form as a direct result of the handling and rolling of the pastry dough the more gentle you are during these stages the better.

It is most important to avoid the pastry sticking either to the work surface, or to the rolling pin. If the dough does stick then it will certainly become badly stretched and a lot of gluten will form. To prevent the pastry sticking you should dust the work surface and the rolling pin with a little flour. It is always preferable to dust the work surface and rolling pin, rather than the pastry itself, since that way only the minimum amount of additional flour will be incorporated in the pastry and it will keep its correct proportions.

To roll the dough, put it on the floured work surface in a ball and bring the rolling pin down on the side of the ball closest to you squashing it down, then roll the pin away from you while maintaining an even pressure on the pin. Then turn the dough through 90° and bring the pin back to the closest part of the dough to you and repeat the exercise until the dough is rolled out to the required thickness. As a general guide for pies, etc. the pastry should have a thickness of between 2 and 4 mm. The actual thickness is really a matter of taste, however, thinner crusts are very likely to puncture during cooking, and if used with pies where a significant amount of steam may be generated during cooking, may start to dissolve away before the pie is finished! On the other hand very thick pie crusts tend not to get fully cooked on the inside and can have a rather sticky finish.

The amount of pastry in the above recipe is sufficient to make a circle of about 30 cm diameter. However, since you will almost certainly not be able to roll it out into a perfect circle do make sure you allow for some wastage. Once the pastry has been rolled out you should leave it for a few minutes to allow any relaxation of stretched molecules to occur before cutting to shape and final use.

What could go wrong when making shortcrust pastry and what to do about it

Problem	Cause	Solution
Pastry dough is very dry and breaks when you try to roll it	Insufficient water has been used.	Very carefully break up the dough with a sharp knife, add a little more water and remix.
Pastry dough is damp and sticky	Too much water was used when making the dough	You may be able to rescue the dough by adding more flour – however, you will need to be very careful to avoid stretching the dough as you mix this extra flour in. The best way is to cut the dough into small pieces using a sharp knife and to coat each piece with flour and then pull the whole mixture back into a ball.

Problem	Cause	Solution
Cooked Pastry is tough	Too much gluten was formed during the mixing and rolling processes	Next time take more care to avoid stretching the dough at all stages
Pastry blisters during cooking	The water was not added evenly	Next time take more care to ensure the water is uniformly distributed through the pastry dough.
Pastry shrinks during cooking	The rolled dough was not allowed to rest for long enough before it was cut to shape – or the shaped and rolled dough was further stretched when finishing.	Next time leave the rolled dough for longer before cutting to shape, avoid stretching the dough to make it fit in a dish, etc.

Making shortcrust pastry using a food processor

If you own a food processor you can use it to rub in the fat. Simply put the flour (and salt if used) in the processor bowl fitted with the sharp metal cutting blade, add the fat, cut into suitable sized pieces, and process at a medium speed for a minute or two. If you do use a food processor the fat will be better rubbed in and you can use a little less fat in the recipe (you can reduce the fat from 50% by weight of the flour to about 40% – i.e. in the above recipe from 150 g to 120 g). However, you will need to use mostly lard, rather than butter as the fat, since there is some water in the butter that can hydrate the starch granules. When you use butter, gluten formation can occur as the starch granules are beaten so vigorously by the processor blade. Also do not be tempted to add the water and use the processor to form the dough – the action of the blade will knead the dough and a great deal of gluten will inevitably form, making for a tough pastry.

Short crust variations

Flan pastry

Flan pastry is very similar to short crust pastry, but uses beaten egg, to replace some or all of the water in the short crust recipe above. The egg proteins, when cooked provide additional binding of the pastry to allow it to withstand the weight of filling with no support. Depending on the flan filling some sugar may also be added to the mixture. Usually flan pastry recipes call for more fat than in the basic recipe above. The actual amount of fat you use is largely a matter of choice, however, as a rough guide it is customary to use 2 parts flour and 1 part fat in basic short crust pastry and to use up to 3 parts flour to 2 parts fat in the richer flan pastries. Apart from the change in the proportions of the ingredients the method of preparing the flan pastry is exactly the same as that for basic shortcrust pastry.

Paté sucrée

This is a rich, sweet variation of shortcrust pastry, used in French pies and tarts. The ingredients are 2 parts flour, 1 part caster sugar and 1 part butter bound together with egg yolks rather than water. You will need about 1 large egg yolk for every 80 g of flour. Again, the method is exactly the same as for the basic shortcrust pastry.

Steak and Kidney Pie (an example of a single crust pie)

Ingredients (for 4 people)

250 g	Short crust pastry (i.e. short crust pastry made using 250 g of flour about 400 g total weight)
300 g	Stewing beef
200 g	Beef kidneys
500 ml	Beef stock (see Chapter 9 for recipe, or use a stock tablet)
25 g	Cornflour
1 egg beaten	

Method

Begin by cooking the steak and kidney; this will take two or more hours. Cut the steak and kidney into rough cubes about 1cm in size. Brown the steak and then the kidney in a frying pan with a little fat, or oil. Make sure the meat is well browned to bring out all the flavour through the Maillard reactions (see recipes in Chapter 6 for details). Put the browned steak and kidney in a saucepan (or casserole dish if you intend to cook the meat in the oven). Deglaze the frying pan with the stock and add to the meat in the saucepan. Simmer for about two hours, checking regularly that the liquid does not boil away. If you have any problems with the meat sticking to the saucepan, then use a casserole dish and cook the meat in the oven at 160°C instead.

While the meat is cooking prepare your pastry according to the recipe above and leave it in the fridge until needed.

When the meat is ready, remove it from the stock and set aside. Make the liquid up to about 400 ml by adding boiling water. Thicken the stock with the cornflour suspended in a little cold water – just add the suspension to the stock and bring to the boil while stirring, set aside.

Put the steak and kidney in a suitable sized ovenproof dish (it should roughly fill the dish), add a little of the thickened stock, enough to cover the bottom of the dish a couple of millimetres deep. It is a good idea to grease the edges of the dish to prevent the pastry sticking. If you wish place a pastry funnel in the middle of the dish – the funnel allows steam to escape from the pie as it cooks so if you do not use a funnel you will need to make some holes in the pastry.

Take the pastry out of the fridge and roll it into a sheet about 1 cm larger than the top of the dish you have used. Pick up the pastry sheet and drape it over the dish. Cut around the edges of the dish with a sharp knife and press the pastry down around the lip of the dish to make a seal. If you are not using a pie funnel cut a few holes in the pastry. Use the pastry trimmings to make some decorations, such as leaves, or the diners' initials, etc. Brush the top of the pastry with the beaten egg and stick on the decorations.

Put the pie in a preheated oven at about 170 °C for about 25 minutes until the pastry is browned. While the pie is cooking reheat the stock and add pepper and salt to taste to make the gravy to serve with the pie.

Fruit Pies (double crust pies)

Ingredients
500 g short crust pastry (i.e. short crust pastry made using 500 g flour, about 800 g total weight)
500 g fruit of your choice (e.g. apples, etc.)
100 g sugar (adjusted to taste)
1 egg beaten

Method

Begin by preparing the pastry (according to the recipe above) and cooking the fruit. The fruit should be cleaned, peeled and cut into 1 cm sized pieces, and then cooked gently in a little water with the sugar added. Try not to over cook the fruit, it should remain firm, but not crunchy. Strain the fruit and keep the liquid it was cooked in. Boil this liquid down until it has a volume of about 20 ml.

Take the pastry from the fridge, divide into two pieces and roll them out so that one piece is about 5 cm larger in diameter than the pie dish and the other is about 2 cm larger than the dish. The dish should be about 20 to 25 cm in diameter – if you use a larger dish you will need more pastry, etc. Line the pie dish with the larger of the two pieces of pastry carefully pressing down into the bottom and sides of the dish to make sure there is no air trapped. Now fill the lined dish with the fruit. Pour some of the liquid over the fruit so that it is just damp, taste for sweetness and if necessary sprinkle some sugar over the fruit. Next brush some of the beaten egg on the pastry around the lip of the dish and put the other piece of pastry on top. Pinch the two pieces of pastry together to seal them and then trim off the excess using a sharp knife. Make several holes in the top of the pie to allow steam to escape as the pie cooks and brush the top with the sugary liquid.

Bake in the oven at about 180 °C for about 20 minutes, until the pastry is golden brown. Serve with egg custard (see Chapter 9 for recipe).

Raised Pie Pastry

Raised pies are a British speciality and are widely available from delicatessens and the better quality supermarkets. However, a home made raised pie is straightforward to make and will greatly impress any guests. The basic difference between raised pie pastry and shortcrust is that you have to make the raised pie pastry strong enough that it will not collapse under the weight of the filling as it cooks. This requirement for physical strength in the pastry means that the starch granules need to be swollen up before the cooking begins. Accordingly the pastry is made with hot, rather than cold, water. Of course the use of hot water which quickly swells the starch granules makes the formation of gluten much more likely and inevitably, the raised pie crust will have a tougher and harder texture than will short crust pastry. The formation of gluten (and the ultimate strength of the pastry) is also promoted by the use of a lower proportion of fat than in other pastries.

Once the pastry has been prepared, it must be formed into the shape required for the particular pie (this formation of the pie case is the 'raising' process). There are several different methods you may use to raise the pie case. You can mould the warm dough around the outside of a suitable container such as a jam jar and allow the dough to cool and harden before carefully removing the pie case. Alternatively you can line a suitable cake tin with the warm dough, in which case you can either cook the pie with the pastry still in the cake tin, or preferably, you can remove the case from the cake tin before cooking – thus ensuring a good brown colour to the outside of the pie case. If you are particularly rich you can buy special pie moulds that allow fancy shaped pies to be moulded (these moulds typically cost in the region of £50, so unless you are going to make a business of pie making they are unlikely to be worth the investment).

Once the pie case has been filled paint the top edges of the pie case with a little beaten egg and roll out a circle for the pie lid and carefully drape it over the, now full, raised pie case – pinch the edges together to seal and make a scalloped pattern. Cut a hole in the middle of the top. Once the pie has cooked and cooled down you can pour some warmed jellied stock through this hole. When this stock cools it will re-form the jelly and fill up any spaces in the pie.

Key points to bear in mind when making raised pie pastry

- Make sure the pie case is well sealed – i.e. there are no holes in the sides
- Allow the pie case to cool and harden before removing it from the mould
- If using a cake tin mould fill the pie case before removing from the tin.
- Always cut a hole in the top of the pie to allow you to add some jellied stock after the pie has been cooked

Why these key points matter

Since you will be using the pie with the crust being self supporting, it is crucial that it is strong enough to withstand the weight of the filling inside and that it does not leak and so wet the outside of the crust (thus weakening it) during the cooking process. However, you need to have a hole in the top of the pie both to allow steam to escape during cooking and also to allow you to pour in some stock that will form a jelly to fill in any voids in the finished pie.

Basic Recipe for Raised Pie Pastry

Ingredients
600 g Plain Flour
15 g Salt
180 g Lard
280 ml water

Method

Mix the flour and salt in a bowl and make a well in the centre. Put the lard and water in a saucepan, heat to melt the lard and then bring the mixture to the boil. Pour the hot liquid into the well in the centre of the flour and quickly mix with a spoon to form a soft dough. Bring the dough together by hand (beware the mixture will still be quite hot so take care not to burn yourself) and knead a little to firm the dough up. Cover and leave the dough to cool a little until you can easily handle it. Next roll the dough into a sheet about 5 mm thick and use this sheet to mould the pie case (leave it to rest for a few minutes before moulding).

To make small pies up to about 8 cm in diameter, take a suitable sized glass or ceramic jar and grease it lightly on the outside. Place the jar upside down on the work surface and drape the rolled pastry over it. Press the dough onto the jar with your hands moulding it into shape. Leave to cool for an hour or so (covered with a clean tea towel), before carefully easing the jar out of the pie case with a twisting motion. The pie case is now ready for filling.

To make larger pies lightly grease the inside of a suitable sized cake tin – preferably one with a loose bottom and a lever arch action. Take the rolled pastry and drape over the tin and carefully press the dough down into the sides of the tin, being very careful not to puncture the pastry, raise the pastry about 5 mm above the cake tin to leave a lip to help secure the top. Leave (while covered with a clean tea towel) to cool and harden for about an hour and then fill the pie. You can then remove the pie case from the cake tin, or you can put the pie lid on and partially cook the pie before removing the tin.

Once the pie case has been filled, roll out some more of the pastry to form a circle suitable for the pie top. Brush the top of the pastry pie case with some beaten egg (to act as a glue for the lid) and drape the lid over the pie case. Pinch the lid and the pie case together between your thumb and forefinger, making a 'scallop' pattern around the pie case. Trim off any excess pastry and cut a hole about 4 mm in diameter in the centre of the pie lid. Finally brush the pie all over with a little beaten egg (to promote the browning of the outside of the pie) and decorate with pastry scraps cut into leaf shapes, etc.

Cook the pie for about 15 to 20 minutes in a hot oven (ca. 220 °C) to help set the pastry and then reduce the temperature to about 180 °C and cook for a fur-

What could go wrong when making raised pies and what to do about it

Problem	Cause	Solution
The pie case collapses under its own weight.	Either the pastry was prepared with too much liquid, or it was not allowed to cool enough before being used, or it was not kneaded enough before use.	Next time use a little less liquid in the dough. For now, knead some more flour into the dough and try to mould it again.
Pastry is too brittle to mould.	Either too little liquid was added to the dough, or it was allowed to cool too much before moulding.	Try gently warming the dough in a covered basin in the oven to soften it enough to make it pliable.
Cooked pastry is too dry	Either the filling itself is so dry that it has absorbed moisture from the pastry, or the pie was cooked for too long a time, or the dough did not have enough liquid added in the first place.	There is little you can do except serve the pie with a sauce. Next time you can either cook for a shorter time, or use a moister dough or a moister filling.
Cracks appear in the case during cooking.	The pastry was not properly moulded, so that in places where there were joins in the pastry it has not properly stuck to itself.	You can prepare a little more pastry and, after coating the cracks with a little beaten egg, fill them with the new pastry which should prevent them from leaking later. Nb you must make this sort of running repair while the pie is still cooking and should allow at least 20 minutes for the added pastry to cook.

ther 50 to 90 minutes until the pie crust is well browned. If the pie is being cooked in a cake tin remove it from the tin at least 30 minutes before the end of the cooking time and brush the sides with a little beaten egg so that the sides are well browned.

Take the cooked pie from the oven and allow to cool to room temperature. Take a little jellied stock (prepare stock as in Chapter 9 and if necessary add a little gelatin to make it set following the gelatin manufacturers instructions) and gently heat it until it just melts. Pour the liquefied stock in to the pie through the hole in the lid and fill it to the top, then place in the fridge for two hours to allow the stock to set before serving the pie.

Pork and Chicken Pie

Ingredients
500 g Belly pork
250 g Chicken Breasts
Fresh Parsley

600 g quantity of raised pie pastry (i. e. use 600 g flour as in the recipe above)
250 ml jellied stock (this will come from the cooking of the meats, but you may need to add a little gelatin)

Method

You will need to cook the meats the day before you prepare the pie. Cut the pork into pieces about 1 cm in size and brown in a frying pan then put it in a saucepan, cover with water (or stock if available) and simmer for about 2 hours. Cut the chicken into strips about 0.5 cm wide, brown in the same frying pan and then put in another saucepan, cover with water or stock and simmer for about 1 hour. When the meats have been cooked, drain off and keep the liquid in which they were cooked and set aside in the refrigerator. Take the liquid and reduce it by boiling to about 300 ml (if there is less liquid then make it up to 300 ml). Allow the liquid to cool and see whether it sets to a soft gel. If the liquid does not gel then warm it over a gentle heat; at the same time dissolve about 20 g of gelatin in a little hot water and add to the liquid and set aside to cool again – this time it should set into a soft gel.

The following day, prepare the meats and parsley for the pie. Take the pieces of pork and gently rub them between your fingers to break them down a little, tear the chicken into thin strips and finely chop the parsley. Prepare the raised pie pastry as described above and mould about 2/3 of the pastry into a pie case about 12 cm in diameter and about 8 cm high.

Fill the pastry case in layers – first a layer of pork, then a layer of chicken and then a layer of parsley, and so on until all the meat is in the pie case. At this point the filling should be about 1 cm below the rim of the pie case at the edges and piled up to about level with the rim in the centre.

Once the pie case has been filled, roll out the remaining pastry to form a circle suitable for the pie top. Brush the top of the pastry pie case with some beaten egg (to act as a glue for the lid) and drape the lid over the pie case. Pinch the lid and the pie case together between your thumb and forefinger, making a 'scallop' pattern around the pie case. Trim off any excess pastry and cut a hole about 4mm in diameter in the centre of the pie lid. Finally brush the pie all over with a little beaten egg (to promote the browning of the outside of the pie) and decorate with pastry scraps cut into leaf shapes, etc.

Cook the pie for about 15 to 20 minutes in a hot oven (ca. 220 °C) to help set the pastry and then reduce the temperature to about 180 °C and cook for a further 50 to 90 minutes until the pie crust is well browned. The crust is cooked when, if you tap the side with your knuckle, it feels hard and sounds solid. If the pie is being cooked in a cake tin remove it from the tin at least 30 minutes before the end of the cooking time and brush the sides with a little beaten egg so that the sides are well browned.

Take the cooked pie from the oven and allow it to cool to room temperature, this will take several hours. Take the jellied stock and gently heat it until it just melts. Pour the liquefied stock in to the pie through the hole in the lid until it starts to overflow, then place in the fridge for at least two hours to allow the stock to set before serving the pie.

Puff Pastry

Puff pastry is often regarded as being very difficult to make. However, most of the apparent difficulties lie in the careful preparation of the ingredients and taking real care with the method. Once mastered, home made puff pastry is a wonderful and most impressive trick.

Basically all puff pastry is just a set of thin sheets of high gluten content pastry that are separated by some fat and when cooked, steam is generated from the fat which separates the leaves of the pastry so that a multi-layered fine textured and crisp product is revealed.

This is one form of pastry where you actually need to ensure a good formation of gluten as the pastry is formed. The basic method is to first prepare a dough with only a comparatively small fat content and then to knead this dough to form gluten sheets within the dough. The kneaded dough is then rolled to form a thick sheet and left to relax all the by now stretched gluten. A large amount of solid fat (mostly butter) is then carefully wrapped in the relaxed pastry sheet. The edges are carefully sealed and the whole packet is rolled out and folded back on its self. The rolling and folding process is then repeated several times (with short resting periods interspersed to allow the gluten to relax) to build up to as many as a thousand or so layers of pastry separated by thin layers of fat.

When cooked the water in the butter turns to steam and pushes the pastry layers apart providing the characteristic flaky texture. Obviously, it is very important to keep the layers of pastry separate from one another and to retain the thin layer of fat between them. As the layers become thinner so there will be a tendency for the fat to become absorbed in the starch above and below it – note that the total contact surface area of the fat with the starch doubles with each rolling so the amount of absorption of fat in the starch also increases. To minimise the extent of this incorporation of the fat in the starch and so keep the layers well separated you need to keep the whole dough as cool as possible – at lower temperatures the absorption of the fat is much reduced. So it is best to carry out the whole process in as cool a room as possible and to store the pastry in a fridge between successive rollings so as to keep it cold.

You will find in many cookery books recipes for a version of puff pastry called flaky pastry. This pastry is basically similar to puff pastry, but instead of carefully putting the butter in a separate piece and rolling several times, you only roughly incorporate the butter, or fat, in the pastry mix, so that it makes lots of small flakes, rather than many fine layers.

Key points to bear in mind when making Puff pastry

- Use at least a proportion of some strong (bread) flour (to increase the gluten content)
- Once the pastry dough is first made knead it for a few minutes to allow the formation of gluten sheets
- Be careful to allow all the internal strains to relax between processes
- Once the fat has been wrapped in the basic pastry be very careful to keep everything as cool as possible
- Keep the pastry a uniform thickness during each rolling
- Keep the pastry square (i.e. make sure the sides are straight and at right angles to one another) when rolling it out so that it will fold up into a neat rectangular shape after each rolling.

Why are these key points important?

It is important to form enough gluten to make the pastry dough strong enough to withstand the rolling into very thin sheets without these sheets breaking up. It is also very important to keep the thicknesses of all the separate pastry and fat layers as uniform as possible. If the layers start to have different thicknesses it is likely that the thinner layers will start to get so thin that they break up. Leaving the rolled dough to relax in a cool place serves two purposes. First it allows the fat (which will be heated by the mechanical energy of rolling) to cool down and remain solid (as long as the fat is solid it is unlikely to penetrate much into the starch granules of the "pastry" layers. Secondly, it allows the dough to recover any elastic deformation that remains after the rolling and thus helps in keeping the dough square.

Basic Puff Pastry Recipe

Ingredients
250 g Plain Flour (at least half should be a strong bread flour with a high protein content – this will help give the elasticity needed to the separate layers)
5 g salt
250 g Butter
125 ml cold water
squeeze of lemon juice (optional) to help the gluten formation.

Method

Rub 40 g of the butter into the flour and salt (either by hand or using a food processor), following the instructions for shortcrust pastry above. Add the water and lemon juice slowly and form into a dough either with a spoon or by hand. The dough should be fairly soft. Knead the dough for a few minutes on a floured board, or work surface, until it becomes stiffer and quite elastic.

Take the rest of the butter and form it into a rectangular piece about 10 by 5 by 2 cm. Leave the butter in the fridge for about 10 minutes to harden up. Form the pastry dough into a rectangle about 24 cm by 12 cm. Place the butter in the middle of the rolled pastry and carefully fold the sides up over the top of the butter and seal by pressing together with your fingertips. Take a rolling pin which has been dusted with flour, and bring it smartly down on one end of the packet so that the thickness is roughly halved from about 6 to about 3 cm. Roll the packet out away from you so that it has a uniform thickness. Bring the rolling pin back to the start and finish rolling to a thickness of 2 cm. Once the dough has been rolled out in this way, fold it in three so that it has roughly its original dimensions and then turn the whole thing through 90 °C and repeat the whole process. Once the pastry has been rolled twice in this manner, set it aside to rest and cool down in the fridge. If you find the fat starts to become soft during the first rolling then put in the fridge to rest and cool between each rolling. After about 30 minutes in the fridge take the dough out and roll and fold it twice more as described above, return the dough to the fridge for another 30 minutes before rolling and folding yet another two times. Finally leave the pastry for about another 30 minutes before using as directed in the particular recipe.

What could go wrong when making puff pastry and what to do about it

Problem	Cause	Solution
The pastry fails to rise	The fat has become mixed in with the pastry and not formed separate layers. The pastry did not contain enough gluten so it could not be rolled into thin sheets. This can happen if the temperature of rolling is too hot Or if the pastry is not left to rest between rollings Or if it has been rolled too much so that the fat layers have become so thin they have mixed with the pastry.	Next time: Work the pastry more to make extra gluten before making the packet with the fat. Make sure the pastry is cooled before rolling (ideally it should be at about 5°C to 10°C) Make sure you leave the pastry to rest for at least 20 minutes between turns Make sure you start with the fat at least 2 cm thick and the pastry at least 1.5 cm thick. Be careful that you never roll the pastry to less than 2 cm thick overall while making the pastry. Otherwise there is a risk the fat layers will become so thin they break up and mix in with the pastry layers.
The pastry rises in some places but not in others.	The fat broke into islands during rolling – the pastry rises where there is fat between the layers and does not rise where there is no fat. This can happen either because you have rolled the pastry unevenly, or if the fat is too cold and fractures rather than flows plastically when rolled.	Be careful to roll the pastry evenly and do not allow the fat to become too cold (it should be at a temperature of 5°C or higher when rolling)

How many layers are there in puff pastry?

When you make puff pastry you begin by making a small packet of fat wrapped in pastry. The fat in the middle is about 2cm thick and the pastry wrapping it is about 1.5 cm thick, making the whole package about 5 cm thick. You then roll this out until it is just a bit less than 2 cm thick and fold it into three. In this new packet we have three layers of fat separated by layers of pastry. The two outer layers of pastry are 0.5 cm thick and the two layers in the middle, between the fat layers are 1 cm thick; the three fat layers are each 2/3 cm thick. So the whole packet is 5 cm thick again.

Again this is rolled out and folded into three. At this stage we have 9 fat layers separated by pastry layers. The two outer pastry layers will now be about 1.7 mm thick and the inner 8 layers are 3.3 mm thick. The fat layers are all about 2.2 mm thick.

You may now be able to see a pattern emerging. After each "turn" in the rolling we multiply the number of fat layers by 3 (and divide their thickness by three) and the number of pastry layers is always one more than the number of fat layers (as both the top and bottom layers are both pastry).

So after n turns, we will have 3^n fat layers and $3^n + 1$ pastry layers. As we usually make 6 turns in total there will be: 729 fat layers (each about 0.025 mm thick) and 730 pastry layers (about 0.04 mm thick in the middle and 0.02 mm thick on the outside).

Sausage rolls

Ingredients
250 g quantity of puff pastry (i.e. puff pastry made with 250 g flour)
250 g sausage meat
A little beaten egg

Prepare the puff pastry as described in the recipe above. Heat the oven to 200 °C. Prepare a baking sheet by sprinkling it with some flour. Divide the sausage meat into three portions and roll each into a cylinder about 2 cm in diameter and about 20 cm long. Roll out the pastry to form a rectangle about 25 cm square with a thickness of about 5 mm. Cut the pastry into 3 strips each about 8 cm wide. Now put one of the sausage meat cylinders along the middle of each of the pastry strips. Brush the beaten egg along the edges of the pastry strips and fold them over the sausage meat and seal them down with your fingers. Cut the rolled up pastry into pieces about 5 cm long and score some lines on the surface to allow steam to escape from the sausage meat during the cooking. Brush a little more of the beaten egg over the tops of the sausage rolls. Place the sausage rolls on the baking tray and cook in the oven at 200 °C for about 15 to 20 minutes until well browned.

Strawberry Slice

Ingredients
250 g quantity of puff pastry (i.e. puff pastry made with 250 g flour)
250 g Fresh Strawberries
250 ml thick egg custard made from 2 egg yolks, 180 ml milk, 50 g sugar (see Chapter 9 for instructions)
A little beaten egg
A little strawberry jam

Prepare the pastry according to the recipe above. Prepare the custard using the ingredients listed here and the method described in Chapter 9. Preheat the oven to 200 °C. Roll out the pastry to form a rectangle about 25 cm by 20 cm with a thickness of about 5 mm. Cut two strips about 2.5 cm wide from the 20 cm side of the pastry and glue them with a little beaten egg back on top of the larger pastry rectangle to form a "tray". This "tray" should have sides about 1 cm high and about 2.5 cm wide and be about 5 mm high in the middle which should be about 10 cm wide. Put the pastry "tray" on a baking sheet and cook in the oven for about 15 minutes during which time it should rise considerably. While the pastry is cooking, wash the strawberries and, if they are large, cut them into pieces about 2 cm in size.

Take the cooked pastry from the oven and allow it to cool a little. Next carefully pour the now cooled and thickened custard in the middle of the pastry case – the custard should be thick enough that it does not flow over the edges. Next arrange the strawberries neatly in the pastry using the custard to hold them in position. Finally, heat the strawberry jam to melt it and brush a little over the tops of the strawberries to glaze them. Allow to cool before serving.

Choux Pastry

Choux pastry is used for eclairs, profiteroles and other filled pastries, both sweet and savoury. The texture should be a little stiffer than that of whipped cream, but still suitable for piping into shapes, etc. When cooked the pastry will rise to leave quite large holes inside – these voids are ideal for filling with sweet, or savoury pastes. The cooked pastry should be crisp but will quickly become soggy and tough on exposure to moisture, so it is best cooked and eaten soon after.

To achieve the rising of the choux pastry, the outside is cooked to form an elastic and deformable film, rather like an un-inflated balloon; then as steam is generated within the pastry mixture it 'blows up' the balloon-like pastries. To make sure that the pastry can expand as the "balloons" are inflated you need to ensure that there is a reasonable amount of the elastic gluten present. Too much gluten will lead to a very stiff balloon that will only expand a little under the internal steam pressure, while too little gluten will lead to rapid expansion and the bursting of the balloons under the pressure of steam generated within the pastries. So once again it is crucial to control the degree of gluten formation as the basic dough is prepared. There is an additional complication when making choux pastry since it is made with eggs. The eggs are needed to provide sufficient protein to bind the outsides together so that a continuous rubber like sheet is formed on the outside of the pastry as it starts to cook. So the amount of egg used will also affect the properties of the rubber like outer layer. Too much egg leads to a stiff and hard outer layer while too little leads to a thin and easily broken layer.

Since the properties of the "rubber like" outer layer depend both on the gluten content and the amount of egg used, you can, to some extent, play off adding more egg against creating less gluten. This interplay between gluten formation and adding extra egg proteins allows some control of the final texture of the cooked pastry. If you rely on a lot of gluten and little egg, the texture will be rather tough, while if you use a great deal of egg protein and form only a little gluten, the texture will be very crumbly – more like a cake than a pastry.

The pastry is made by melting the fat in the water and then quickly adding the flour – this fast addition of the flour is important since the hot water is very quickly absorbed by the flour and you want to ensure that the uptake of water is uniform throughout the paste. If the flour is added slowly the first flour to be in the water would take up more moisture than the rest and the result would be a non-uniform dough that would cook unevenly.

Key points to bear in mind when making choux pastry

- Add all the flour very quickly to the hot water and melted fat
- Beat in the eggs gradually making sure that the mixture has the 'right' consistency
- When cooked split open pastries that are to be filled so that the insides are dried and crisp

Basic Recipe for Choux pastry

Ingredients
125 g Plain flour
50 g Butter
150 ml water
2 eggs (beaten)

Method

Melt the butter in the water and bring to the boil. Note, it is important not to bring the water to the boil before the butter has melted since the boiling water will evaporate while you wait for the butter to melt and the proportion of water in the recipe will be reduced. As soon as the water is boiling remove from the heat and tip in all the flour at once. Quickly beat the mixture to form a smooth paste – do not beat any more than necessary at this stage. The fat and the water are both absorbed into the starch, however, if you continue to beat the mixture some of the fat may be released and replaced by water so the mixture will become 'fatty' on the outside and the extra absorption of water may lead to the formation of to much gluten.

Allow the mixture to cool for a few minutes before beating in the eggs. The temperature should be low enough that the egg proteins do not coagulate when added to the mixture (i.e. below about 60°C). Now beat in the eggs adding a little at a time. You should aim to incorporate some air during this beating – the air will form small bubbles that will act as nucleating sites for the larger steam bubbles that will form during cooking. The more small air bubbles you manage to incorporate, the more even the final internal texture will be. You should continue to beat the mixture until it has the stiffness and texture of well whipped cream. A 'sheen' should develop as the beating continues – once this sheen has appeared stop beating and use the paste as required by the recipe.

What could go wrong when making choux pastry and what to do about it

Problem	Cause	Solution
Pastry does not rise	Either no air was incorporated into the pastry, or the egg proteins coagulated and "cooked" before the pastry was put in the oven	Make sure the pastry has cooled to well below 60°C before adding the eggs (otherwise they can "cook" during the mixing) and make sure you beat in plenty of air in the mixing.
Pastry does not become crisp	The mixture was too moist, or the oven was not hot enough, or the cooking time was too short	If the pastry has browned on the outside, then it is most likely too moist a mixture, so add less liquid next time. If the pastry is not browned it is probably that the oven temperature is too low or the cooking time too short – try increasing the temperature and/or the cooking time.
When split open there are no "holes" inside to fill with cream, etc.	Insufficient air was incorporated in the mixture, and/or the mixture was too stiff	Beat plenty of air in the mixture and make sure it is still very soft before using.

Profiteroles

Ingredients
Choux Pastry made using 125 g Plain flour, 50 g Butter, 150 ml water, 2 eggs (beaten)
150 ml whipping cream
200 ml Chocolate Sauce made using 150g dark chocolate, 50 ml milk

Method

Heat the oven to 250°C and prepare a baking sheet with a sprinkling of flour. Prepare the Choux pastry as described in the recipe above. Put the choux pastry in a piping bag with a wide (1 cm) plain nozzle and pipe into about 12 small drops about 2 cm across on the prepared baking sheet. Cook the choux pastry for 15 to 20 minutes until it is crisp and lightly browned on the outside.

Allow the profiteroles to cool and split them open with a sharp knife. Whip the cream until it is stiff and fill each profiterole with a spoonful of cream and then close them up to their original shape, so the cream is not visible.

Prepare the chocolate sauce by breaking the chocolate into pieces and gently warming with the milk stirring all the time until the chocolate has all melted. Pile the profiteroles on a plate and pour the warm chocolate sauce over them, serve immediately before the chocolate sauce sets.

Cheese beignets

Ingredients
1 egg quantity of Choux pastry using 60 g Plain flour, 25 g Butter, 75 ml water,
1 egg (beaten)
120 g Gruyere Cheese – grated
30 g grated Parmesan Cheese

Method
Prepare the choux pastry as in the recipe above, stir in the grated Gruyere cheese and roll into 8 small balls about 2 cm in diameter. Heat some clean cooking oil
in a pan to 180 °C and deep fry the beignets 3 or 4 at a time for 3 to 5 minutes
until they are well browned on the outside. Lift the cooked beignets out of the
oil, drain on kitchen tissue and then roll in the grated Parmesan cheese to cover.
Serve while still hot.

Some experiments to try at home

An experiment to illustrate the number of layers in puff pastry

This is a simple experiment that will help to illustrate the exponential increase
in the number of layers as puff pastry is rolled out.

Simply take two differently coloured pieces of plasticine and roll each out to
a square about 1 cm thick and about 5 cm on each side. Put one piece on top of
the other and roll in one direction until the length has doubled (and the thickness halved). Now fold length-wise in half to reform the original square shape
and roll again this time in a direction at right angles to the original direction and
fold in half again. Cut a thin slice off one side of the plasticine and count the
layers. Now repeat the same procedure and again cut off a slice and count the
layers. Finally repeat a third time – this time it is very unlikely you will be able
to count the layers, or even resolve them.

You should now realise just how thin the layers of pastry and fat become when
you roll puff pastry.

An experiment to illustrate the relaxation of the gluten after rolling pastry

When you roll out a piece of pastry the gluten molecules become stretched out
along the direction of rolling. A similar, although altogether much more severe
stretching of molecules occurs in the manufacture of plastic packaging. You can
see just how much extension is locked into most plastic packaging with a simple
experiment.

You will need an empty crisp packet for this experiment. Heat your oven to about 180 °C – if you have a gas oven be sure to turn the gas off once it is hot. Simply take the packet and put it on a piece of aluminium foil and put it in the hot oven. As the polypropylene in the packaging heats up to close to its melting temperature, the molecules that were stretched out in the initial processing to make the packaging film are able to relax. As these molecules move back to their original, random, shapes so the crisp packet will shrink to about one quarter of its size.

You can also shrink yoghurt pots and all sorts of packaging in this way – it is quite interesting to see how the writing on the packaging shrinks with the packets.

In pastry, if you do not allow the gluten to relax slowly at room temperature, or in the refrigerator, after rolling, then it will relax quickly as it heats up in the oven and so shrink away from the sides of our pie dish, etc.

12 Soufflés

A hot, well risen soufflé, fresh from the oven always impresses dinner guests. However, many people think soufflés are difficult to cook. Perhaps this belief stems from some recipes that are more likely to result in disaster than in a beautifully risen, stable soufflé. The basis of all soufflés is a foam made from beaten egg whites. During cooking some steam is generated and the air in the foam expands causing the soufflé to rise. As the egg whites cook so the foam sets. All you need to be able to make perfect soufflés is to understand what ingredients and processes will tend to make the egg white foam collapse, and to avoid using them as far as possible.

Basic Principles

Following a few simple rules will ensure perfect well risen soufflés every time. Begin by preheating the oven and greasing the dishes (try to use smaller dishes where possible).

It is important to put soufflés into a hot oven so that they start cooking quickly. The egg-white foam at the heart of soufflés has a tendency to collapse after the final stages of mixing when any fats are added (as described later) so the sooner it starts to cook the better. It is also important to have the oven at the right temperature; if the temperature is too high then the soufflé may burn on the outside before it is cooked in the centre. Cooking at too low a temperature, however, can mean that the soufflé does not rise very well, as the egg proteins cook and stiffen the foam before enough steam is generated to make the soufflé rise.

Greasing the dishes is essential to allow the soufflé to rise properly. Any hard fat (butter, lard etc.) is suitable – hard fats are preferable to soft ones as they are less likely to flow away during cooking. Egg whites are notorious for sticking to surfaces during cooking. This is because the egg proteins react with some metallic atoms in the glaze. (See Chapters 2 and 5 for more detailed explanations). Greasing the dish prevents the protein molecules reaching the glazed surface and thus reduces the risk of the soufflé sticking. Cleaning off any excess mixture from the rims of the dishes is a good idea, since this mixture would cook first and may stick to the rims so preventing the soufflé rising. A soufflé dish should have smooth vertical sides so that the soufflé does not change its shape as it rises. It is easier to cook good soufflés in small dishes. The foam that makes the basis of the soufflé is a very poor conductor of heat, just like the insulating foam used in cavity walls or in loft insulation. The outside of the soufflé gets hot quickly while the centre takes longer to cook.

The smaller the dishes you use the easier it is to produce a soufflé that is uniformly cooked all the way through to the centre, without the risk of burning the outside. If you use larger dishes it is preferable to cook the soufflés for longer times at lower temperatures; the table on page 202 gives a guide to cooking times and temperatures for soufflés cooked in different sized dishes. Thus it is often preferable to make several individual soufflés, rather than one large one.

Separate the eggs and beat the whites to make a very stiff foam

The most important instruction is to beat the egg whites into a really stiff foam. Without a really firm foam as the basis the soufflé simply will not work! Beaten egg whites provide bubbles that expand to make the soufflé rise and trap the air to make the soufflé light. The smaller the bubbles the better, as really small bubbles (so small that they can not be seen) provide a uniform and smooth texture to the soufflé. The more you beat the egg whites and the whiter they get, the smaller the bubbles are becoming.

Cookery books give a lot of different and sometimes contradictory instructions on how to achieve a stiff foam. The main reasons lie in changes that have occurred over the last few years. Power whisks take the hard work out of beating eggs so there is no need to employ any of the shortcuts offered in many older books. These short cuts aimed at making beating easier; all have good scientific explanations but are only necessary if beating by hand. Examples are: beating over a pan of hot water (which makes it easier to denature egg white proteins – see Chapter 10 for a detailed explanation), using a copper bowl (copper ions can 'cross-link' protein molecules together to make the foam stiffer – see Chapter 2 for a detailed explanation), adding cream of tartar (an acid environment makes proteins easier to denature – see Chapter 2 for a detailed explanation).

When beating egg whites use a metal, hard plastic, or Pyrex bowl in which to beat the eggs

One modern innovation that can cause problems is the plastic bowl. The softer type of bowls (especially the older ones), made from polypropylene or polythene can contain some fat like molecules which, although harmless in themselves, can act like other fats to prevent, or reduce, foam formation.

When choosing the ingredients for the soufflé flavouring avoid using fats as far as possible

Avoiding fats is important since they tend to make the egg white foam collapse. We are all familiar with the foams made by soaps in, for example, washing up liquids or bubble baths. We also know that adding oils (or fats) causes the bubbles to disappear. When you beat egg whites you denature, or alter the structure of, the protein molecules so that they act in a way directly analogous to soaps. The bubbles formed by these denatured proteins when egg whites are beaten are similarly burst by any added oils or fats.

Since foams are destroyed by the addition of fats or oils (see Chapter 2), an important item to look for in a list of ingredients for a soufflé is anything that contains fat. Avoiding fats is the simplest way to ensure a perfect soufflé. This includes the fats that are contained in the egg yolk; that is why, contrary to nearly all recipes I have seen in cookery books, I do not recommend adding the egg yolks back into the soufflé. If you do add the egg yolks they will start to collapse the foam, so that the soufflé will not rise as well, or may not rise at all.

Prepare the filling as a very firm paste and fold into the beaten egg whites

The reason the recipe below calls for the apple purée to be very stiff is that this filling provides the strength for the cooked soufflé enabling it to support its own weight. A (flavourless) soufflé can be made with beaten egg whites alone. However, apart from the flavour (or lack of it!), such a soufflé is rather weak and can collapse under its own weight when taken out of the oven. The filling therefore has to provide some additional strength to counter this weakness. Most recipes call for you to use a flour based paste to strengthen the soufflé; however, I prefer to keep the amount of added flour to a minimum as I believe it can affect the taste of the soufflé. Providing you prepare a very stiff and thick purée for the filling you should not need to use any flour.

Dish diameter (cm)	Dish depth (cm)	Oven Temperature (°C)	Cooking Time (minutes)
6	3	180	10
10	4	180	15
10	5	170	20
15	5	160	25
20	5	160	30

Key Points to consider when making soufflés

- Preheat the oven
- Grease the dishes
- Separate the eggs and beat the whites to a very stiff foam
- Avoid using fats in the flavouring, as far as possible

Why these key points are important

To make a good soufflé you need to begin by making the bubbles that will keep the soufflé light. The bubbles are prepared by beating egg whites; whilst it is possible to beat eggs which have not been separated (see the recipe in Chapter 10 for a Genoese Sponge) the fats in the yolks tend to make this difficult and reduce the stability of the resulting foam. Once you have made your foam you want it to rise in the oven. It will rise when the air in the bubbles expands and some steam is generated inside the soufflé as it gets hot. To make sure the soufflé rises evenly, you need to make sure the egg proteins do not stick to the sides of the soufflé dish – this is easily achieved with a little grease on the insides of the dish. Finally, to prevent any risk of collapse of the carefully prepared foam, you should try to avoid using fats in the filling and start to cook it quickly as possible.

Cooking a soufflé with the oven open

Provided you take enough care in preparing the soufflé mixture it is actually possible, but not recommended, to cook a soufflé with the oven door open. All that happens is that the temperature inside the oven is quite a bit cooler than usual – so you set it on a higher temperature, and the front of the soufflé (nearest the front of the oven) is not as quickly cooked as the back – so you need to turn it round during the cooking.

Since this recipe is so robust, you can be quite rough with the cooking soufflés. For example, you can if you wish open the oven door while the soufflé is cooking and then slam it shut – something that all normal cookery books say will ruin a soufflé. Indeed, when ever I cook a soufflé in a public lecture I invariably open the oven door well before the soufflé is cooked and say – "it still needs a few minutes yet" before slamming the door shut – just to emphasise the point that provided you have made a good stable foam in the first place nothing can go wrong when you cook it. On one occasion, at the Edinburgh Science Festival, my partner was sitting in the audience next to the very distinguished lady who was chairing the session when I did this. She leant over to my partner and said "Oh dear! What a pity the soufflé will be ruined!". She point blank refused to believe my partner's protestations that my actions were quite deliberate and the soufflé would come out just fine, as of course it did.

Apple soufflé

The main example recipe is for a dessert soufflé. Variations are given later that include a range of savoury soufflés suitable for starters and main courses. In order to avoid any fats, the recipe calls for egg whites, but not for egg yolks. The egg yolks are best used to make a custard to serve with the soufflé (see Chapter 9 for a recipe for egg custard).

The recipe is designed to produce a little more than 1.2 litres of soufflé mixture, enough to make 4 individual soufflés in soufflé dishes with a diameter of about 10 cm and a depth of about 4 cm.

Initial Preparation

Preheat the oven to the cooking temperature, see the table above for cooking temperatures and times. Thoroughly grease individual soufflé dishes with butter or any other hard fat.

Ingredients
200 ml   egg whites (about 5 large eggs)
100 g   icing sugar
250 ml   apple purée (made from 400 g apples) thickened with a little flour
               or cornflour if necessary

Method

Begin by preparing the fruit purée, this can be done in advance well before the meal. Cook the fruit in a little water and drain before liquidising the cooked fruit in a food processor or liquidiser. The purée should be very dry and stiff, a good guide is that a spoonful of the purée put on a plate should retain its shape and not flow away. If the purée is even slightly runny you can easily thicken it by adding a teaspoon of cornflour and heating slowly until boiling.

Whisk the egg whites until they are stiff. A guide to when they are ready (if you dare try) is that you can turn the bowl upside down without the egg whites falling out! Add the icing sugar and beat the egg whites some more. Taste a little of the egg white foam to check for sweetness, and add more sugar if needed. Once the sweetness is to your liking then finish off with a little more beating to give a very stiff and glossy finish. Next fold some of the beaten egg whites into the fruit purée to give it a less stiff texture, keep adding more beaten egg white and folding it in until it is all used up.

Immediately, fill the soufflé dishes right to the top, making sure that there are no air pockets by knocking the dishes on a hard surface. Clean off any excess mixture and flatten off the tops with a palette knife. Make sure that there is no mixture over the rims of the dishes as this will cook first and prevent the even rising of the soufflés. Dust the tops of the soufflés with icing sugar and put them in the oven to cook for about 15 minutes until the tops and sides just start to brown. Remember to leave space above the soufflés for them to rise. The soufflés should rise up so that the height is more or less doubled. An interesting finish can be achieved by branding a pattern in the sugar on the tops of the cooked soufflés with a red hot wire or skewer. Serve immediately, before the soufflés start to cool and shrink.

Problem solving with soufflés

Unfortunately there is no way to rescue most failed soufflés once they have been cooked, so the guide below just shows you how to do better next time.

Problem	Cause	Solution
Soufflé does not rise	The bubbles in the egg foam burst before cooking started	
	Insufficient 'soap like' molecules to stabilise foam	Beat egg whites more, make sure there is no yolk in egg whites
	Too much fat or oil in flavourings	Avoid using fatty or oily fillings
Soufflé rises, but it is all air inside	Bubbles in the egg foam burst during cooking, after the top has cooked	
	Insufficient 'soap like' molecules to stabilise foam	Beat egg whites more, make sure there is no yolk in egg whites
	Too much fat or oil in flavourings	Avoid using fatty or oily fillings
Soufflé collapses as soon as it is taken out of oven	Cooked foam unable to support its own weight, insufficient cross-linking of egg white proteins or the filling is not having a reinforcing effect.	
	Not cooked long enough.	Increase cooking time and/or temperature;
	Not enough filling.	Use more filling;
	Filling too runny.	Use drier filling, or thicken with cornflour.

Problem	Cause	Solution
Soufflé rises at first in the oven but collapses during cooking	The bubbles burst when the steam pressure becomes too great, the foam does not cook quickly enough to become strong before the steam is generated.	
	Egg whites not beaten enough to provide small bubbles (small bubbles withstand higher pressures).	Beat eggs longer
	Filling too runny.	Use drier filling, or thicken with flour or cornflour.
Top browned but inside raw	Oven too hot so that top cooks too quickly	Reduce oven temperature, or cook in smaller dishes
Soufflé has chewy texture	Oven too cool, middle overcooked before outside browned	Increase oven temperature, or cook in larger dish
Top of soufflé splits or soufflé rises unevenly	Soufflé prevented from rising by sticking to sides of dish	Make sure dish is well greased
Uneven sides to soufflé	Mixture not filling dish properly so that some large air pockets are left behind	Make sure you knock the dishes on a hard surface to get rid of any large air bubbles before putting in oven

Common Problems with soufflés

Soufflés collapse for many reasons. The most usual causes are: too much fat making the egg white foam collapse; not heating the egg whites enough to make a really stiff foam; using too runny a filling so that it cannot reinforce the egg foam; not cooking the soufflé long enough. Soufflés, even when they have risen properly can sometimes have a rough appearance. The usual cause is that the top of the soufflé was not smoothed off well enough before it was put in the oven.

Often soufflés rise more in the middle than at the edges and the tops break open. This happens when the soufflé dish is not properly greased and the soufflé mixture sticks to the sides and so prevents the soufflé from rising properly.

Occasionally, a soufflé will only rise from one side which results in a lopsided finish. There are two possible causes: some of the mixture has stuck at one place on the rim of the dish where it was not cleaned away properly before putting in the oven or the soufflé has been cooked in an oven with a heating element at the back – the more risen side would have been closer to the heat and would have risen more quickly leading to the lopsided finish.

Soufflé recipe variations

It is instructive to look at some of the worst recipes to be found in cook books. I have found several recipes by reputable authors that are, without the utmost care and greatest skill and experience, almost certain to fail.

The commonest problem is the use of more fat than is strictly necessary for the soufflé. For example, many recipes start with instructions to prepare a roux from melted butter, flour (and in some cases milk as well) and then proceed to use this roux as the basis of a white sauce to which the beaten egg yolks and flavourings are added. This sauce is then folded in to the beaten egg whites. These recipes are adding far more fat than is necessary. The fat in the egg yolks and the fat from the butter (and milk) in the roux will begin to burst some of the bubbles in the egg white foam as soon as they are folded in. The result will be larger bubbles in the soufflé, giving a coarser texture, and a weaker, less well risen product that is more prone to collapse.

Other instructions that add unnecessary complications include cooking with the soufflé dishes on a hot metal plate. If the dishes are placed on a hot plate the soufflés will cook more quickly at the bottom than elsewhere. The result will be a soufflé that is either burnt at the bottom, or else undercooked about half way down. Whenever you are cooking something with a foam structure (such as a soufflé, or a sponge cake) you must bear in mind that it is a good thermal insulator, so that it will only heat through slowly. It is best to try to heat it as evenly as possible, so the use of a hot plate underneath is not to be recommended.

Of course, for certain soufflés some fat is essential as it forms a part of the basic flavour. The most important examples are cheese soufflés and chocolate soufflés. There are several tricks you can employ to reduce the difficulties caused by the presence of the fats in these flavours.

You can avoid the problem of fats in cheese by using a coarsely grated hard cheese, so that the cheese only starts to melt and release its fat once the egg whites in the foam have started to cook and harden. For a chocolate soufflé you can avoid the use of fat altogether by using cocoa powder, rather than whole chocolate (which contains 40 to 50% fat), to provide the flavour.

However, neither of these solutions is completely satisfactory, you may wish to make a soufflé from a soft cheese, and you may want the full richness that comes from whole chocolate in your chocolate soufflé. The trick to be employed in these cases is to encapsulate the fat in a thick starch based sauce so that it doesn't reach the egg white proteins at all, and hence does not cause any collapse of the soufflé.

Encapsulation of fats can be achieved by several means. The easiest way is to use a very stiff, almost solid, starch thickened sauce to surround the fat. A mixture of cocoa powder and cornflour is very suitable for chocolate, while cornflour on its own is acceptable for cheeses. There are two methods that can be employed, either add the solid fatty ingredient in a finely grated form to a nearly set sauce, or add the fatty food to the hot sauce and beat really hard while the sauce cools to divide the fat into fine droplets that solidify and are coated by the sauce.

Chocolate Soufflé

You can apply the first of the encapsulation methods described above to make a simple chocolate soufflé. Begin by making the sauce base by adding a mixture of equal quantities of cocoa powder and cornflour to a little water and mixing to a smooth paste (you should use about 40 g of cocoa powder and 100 ml water for a 4 egg whites soufflé). Heat this paste until it thickens and allow to simmer for a few moments. Allow the thick mixture to cool (below the melting point of the chocolate) and then add about 150 g of finely grated chocolate. Fold the mixture into beaten egg whites as in the recipe above and cook in the usual way. In the soufflé, the egg white will start to cook before the fats are released from the starchy paste so producing an excellent soufflé.

Brie Soufflé

The second encapsulation method can be applied to make a Brie soufflé. Melt about 200 g Brie with a little (50 ml) water, once it is all melted add 40 g cornflour made into a paste with a little cold water. Beat vigorously and continue to heat until the mixture thickens (and starts to bubble). Remove from the heat and keep beating (in a food processor or with a power whisk if available) until the mixture is cool and the fat has set. Fold this mixture into the beaten egg whites as above and cook in the usual way. In the soufflé, the egg whites will cook before any fats are released so producing an impressive Brie soufflé. Of course this recipe can be adapted for use with any cheese.

Unusual recipes that can be made using the science principles

With a good understanding of what makes the soufflé recipes work, you should be able to make a soufflé without any recipe at all. Such a skill might be useful when on holiday where there are, say, lots of cheap plums just waiting to be turned into a delicious dessert, but you have no recipe book to look up what is needed. The confidence of knowing what will work means that you can just make it up as you go along.

Once you have grasped the important principles involved in cooking soufflés, you should be able to move on to using them to make interesting and different types of soufflé. In practice the only limitations to the possibilities are your own imagination, and the problems that are associated with making soufflés with particularly fatty fillings. To start you off thinking up novel soufflés I have put together a few oddball recipes below. In these recipes I have deliberately kept the detail to a minimum, so that you can use your own skills to interpret and adapt them to your own taste.

Spotted Dick Soufflé

Ingredients

200 ml	Egg whites (from about 4 large eggs)
100 g	Icing sugar
150 g	Raisins (and/or sultanas etc.)
30 g	Cornflour

Method

Prepare the soufflé dishes and preheat the oven. Beat the egg whites to a stiff foam and fold in the icing sugar, beat a little more until the mixture is so stiff that you can turn the bowl upside down without dropping the foam on the floor. Fold in the cornflour and gently mix in the dried fruit. Fill the soufflé bowls to the brim, level off the tops with a palette knife, clean off any excess mixture and cook in the usual way. Serve as soon as you take the soufflés out of the oven. N.B. cornflour is used in this recipe to thicken the egg foam enough to make sure that the raisins don't sink to the bottom during cooking.

Smartie Soufflé

A simple variation on the above recipe using Smarties instead of raisins makes for a fun dessert for youngsters of all ages (at least my partner thinks its fun!)

Ingredients

200 ml	Egg whites (from about 4 large eggs)
100 g	Icing sugar
2	tubes of Smarties
20 g	Cornflour
10 g	Cocoa powder

Method

Prepare the soufflé dishes and preheat the oven. Beat the egg whites to a stiff foam and fold in the icing sugar, beat a little more until the mixture is so stiff that you can turn the bowl upside down without dropping the foam on the floor. Fold in the cornflour and cocoa powder and gently mix in the Smarties. Fill the soufflé bowls to the brim, level off the tops with a palette knife, clean off any excess mixture and cook in the usual way. Serve as soon as you take the soufflés out of the oven. Note the colour from the coating of the smarties will diffuse into the soufflé, so if you do not use any cocoa powder (which gives a dark brown soufflé), the finished soufflé will have a rainbow appearance.

Tricolour soufflé

Ingredients

200 ml	Egg whites (from about 4 large eggs)
100 g	Icing sugar
10 g	Cornflour
100 ml	Very stiff Strawberry purée (raspberry or cherry will work as well) – add a little cornflour if necessary
100 ml	Very stiff Blueberry purée (plums will also work) – add a little cornflour if necessary

Method

Prepare the soufflé dishes and preheat the oven. Beat the egg whites to a stiff foam and fold in the icing sugar, beat a little more until the mixture is so stiff that you can turn the bowl upside down without dropping the foam on the floor. Divide into three equal quantities in separate bowls. Fold in the Strawberry purée in one bowl and the blueberry in another, fold the cornflour into the third (the cornflour will act to stiffen the egg-white foam without giving it any colour). One third fill the bowls with the red, strawberry, soufflé mixture and smooth the surfaces with the back of a teaspoon, pour on the white soufflé mix until the dishes are two thirds full, smooth the surface and finish off with the blue soufflé mix, filling the soufflé bowls to the brim. Level off the tops with a palette knife, clean off any excess mixture and cook in the usual way. Serve as soon as you take the soufflés out of the oven.

An Experiment to try for yourself

People sized bubbles

As you have learnt, making soufflés is all about making good stable foams, which are just collections of bubbles. So it is fun to apply the same principles to make really big bubbles – big enough for you to get inside them!

First you need to make a really good bubble mixture. At the heart of the mixture are the soap molecules. You need to get the right ones – these are usually those in washing up liquids or detergents for washing cars. In most washing up liquids the manufacturers add salt to make the liquid thicker. While this may give the impression you are getting more for your money, it is just a marketing ploy. However, the salt in these washing liquids can limit the size of bubbles you can produce from them. So look for a cheap and runny liquid.

To make up the bubble solution take about 4 parts of detergent to about 20 parts water and add about 1 part of glycerine (available from chemists) and mix well. Test the mixture by creating a large bubble with a clean metal coat hanger bent into a roughly circular hoop. Dip the coat hanger in the mixture and lift it out. A thin bubble film should be suspended in the hanger. Then hold the hanger at arms length and swing it around in a circle – a big bubble should be formed behind the hanger.

If no film is formed on the hanger when you take it out of the bubble mixture then add some more detergent. If the film does not form a good sized bubble when you swing the hanger around then add a little more glycerine. Keep adjusting the mixture until you can make good bubbles with the coat hanger every time.

Once you have prepared a really good mixture you are ready to try making giant bubbles. You will need to make up a large loop to make the big bubbles. Take a strip of cotton fabric about 5cm wide and about 3metres long and roll it into a stiff tube about 1cm in diameter. Stitch the tube along its length to hold it together and then join the ends neatly to form a loop. Now sew on another length of cotton fabric to act as a handle. Dip the loop of cotton fabric in the bubble mixture and then take it out and run along holding it out beside you by the handle you made – make sure it does not touch the ground. With a little practice you will be able to get the loop to open up to a ring about 1 metre in diameter and then make a bubble about 1 metre in diameter and maybe as much a 3 metres long.

You can also make a bubble large enough to stand inside using an old hula hoop covered with cotton fabric. You will need a friend to help with this bubble. Put the bubble mixture in a child's paddling pool and stand in the pool yourself inside the hula hoop which should be fully immersed in the bubble mixture. Get your friend to take the hula hoop and quickly lift it up over your head, without touching you at all. This will make a bubble with you inside!

13 Cooking with Chocolate

This chapter is a little different from the previous ones in that I shall concentrate more on the history and production of chocolate, than on its uses in cookery. This is because few people seem to appreciate the complexity of processing involved in the manufacture of chocolate, nor do many of us appreciate the long and fascinating history behind the development of this now common foodstuff.

History of chocolate

Chocolate originated in South America, little is known of the early history of chocolate. As far as we know the cocoa tree first appeared in South American tropical rain forests. Later the tree seems to have been carried Northward towards Mexico probably by the Mayas around 700 AD. The cocoa tree became cultivated throughout the region. Most of the indigenous races Mayas, Aztecs and Toltecs used the beans as food in different ways. Since, as we shall see, the beans contain fat, starch and protein, they probably formed an important part of the diet. There is evidence that the beans were also exported northward, from Central America, into what is now the USA, where they were often so highly valued that some native tribes used them as currency.

The detailed (and known) history of chocolate dates from the voyages of the Spanish to South America. The first cocoa beans were brought to Europe in 1502, after Columbus's 4th voyage. However, at that time, the Europeans had no understanding of how to use the beans. It was not until the conquest of the Aztecs by Cortez later in the century that the Spanish learned how to use the beans in cooking. The commonest use was in the form of a drink made from the roasted beans that were simmered in water, flavoured with hot red peppers and vanilla, and thickened with ground corn.

The word chocolate is derived from the Aztecs names for the tree, and for the drink they prepared from the beans. These words live on in Mexican today as 'choclatl' for the drink and 'cacauatl' for the tree. The Aztecs also used the ground beans as a spice to flavour their meat dishes. These sauces have been developed in Mexico and form the basis of some of the richest dishes of modern Mexican cuisine – see the recipe for "mole" later in this chapter.

In Europe the term "chocolate" was originally applied to a drink similar to the Aztec version, but made using less hot pepper, as the Aztec drink was too spicy for the European taste. Hot foods were still very rare in Europe at that time. Cortez introduced the drink to Spain after returning from his Mexican expedition (1519–28). The chocolate drink was modified and developed in Spain eventually becoming a sweetened drink prepared from the cocoa beans together with milk, sugar and eggs. This Europeanised version of the chocolate drink gradually spread from Spain through Europe and arrived in England in the 1650s, where it became particularly popular. Rather than working with the beans themselves, chocolate was prepared by grinding the roast beans into a solid mass that could be moulded into ingots (nowadays known as cocoa mass) that could be used as and when needed.

By the late 17th century, chocolate houses had become social meeting places all over Europe. Two of the most famous chocolate houses in London were the Cocoa Tree where members of the Tory party met (and some suggest was the de facto headquarters of the party), and White's (opened in 1693) which was a favourite of the Whigs as well as the literary set and gamblers. Only the rich could afford to drink chocolate (the cost of a pound of chocolate mass needed to make the drink was 10 to 15 shillings in 1657, the equivalent today of around £500 per pound of chocolate!). Nevertheless chocolate houses quickly became regarded as rather risqué places to be seen. By the 1750s the general reputation of all chocolate houses was rather akin to that of brothels; indeed two of the scenes from Hogarth's paintings 'The Rakes Progress' (1735) are set in and just outside White's.

The church was strongly opposed to chocolate and at various times tried to ban its use in Europe. One well documented example occurred in the city of Chiapa Real in Spain. The local women had their maids bring them chocolate during Mass. When the bishop, who disapproved of chocolate, tried to stop the practice, a riot ensued in which swords were drawn in the Church. The problem was only resolved when the bishop was suddenly (and suspiciously) taken ill and died. The general opinion was that he had drunk a poisoned cup of chocolate.

The recipes for chocolate drinks were continually developed throughout the 17th and 18th centuries. The drink that was served in White's in the 18th century was made using milk, sugar and the chocolate mass made from ground cocoa beans; it would have been very similar to today's drinking chocolate, except for a high fat content. Indeed the fat which was always present and usually floating on the chocolate was generally disliked and many efforts were made to remove this fat from the beans.

The first report of the use of a press to remove fat from the beans is in a French treatise published in 1678. However, the credit for the invention of the process for extracting cocoa butter from the cocoa bean to make cocoa powder is usually given to the Dutchman, Conrad van Houten. In 1828 he made 'chocolate powder' by squeezing most of the fat from finely ground cacao beans. The fat, which was initially regarded as a waste product, became known as cocoa butter.

The availability of cocoa butter soon led to the appearance of modern chocolate. The first 'Eating Chocolate' was introduced by the Bristol firm of Fry and Sons in 1847. The cocoa butter from pressing was added to a powder-sugar mixture, and the new product was born. In 1876, a Swiss firm added condensed milk to chocolate, producing the world's first milk chocolate. Since then the production and consumption of chocolate has continued to rise every year.

The ever growing demand for chocolate led to the planting of cocoas trees all over the world, wherever the climate is suitable; thus today cocoa beans can come Africa and Asia as well as South America.

From the three chests of cacao beans that Cortez exported to Spain, cocoa bean exports in the world reached over 2 million tons in 1998. One fifth of all the exports went to the United States. Even with this tremendous usage of chocolate, the United States still ranks only tenth in the world with a per capita consumption of 4.5 kg annually, far behind the first-place Swiss, who eat 9.5 kg per person annually. The British take second place with an annual consumption of around 8.8 kg per person.

The History of Chocolate

700	Evidence of cultivation of cocoa trees in South America
1000	Evidence of use of cocoa beans in Mexico
1325	Aztecs drinking chocolate in Mexico
1502	Columbus brings cocoa beans to Spain
1519	Cortez conquers Aztecs
1528	Cortez brings back cocoa beans and recipes for the drink chocolate
1580	First 'Chocolate factories' to process beans set up in Spain
1620	Chocolate described in France as 'the damnable agent of necromancers and sorcerers'
1648	Bishop of Chiapa Real murdered for trying to ban chocolate
1650	Chocolate arrives in England
1693	Whites chocolate house opens in London
1735	Hogarth depicts Whites in his series of paintings, 'The Rake's Progress', as a den of iniquity
1828	van Houten invents cocoa press
1847	Fry and Sons produce the first eating chocolate
1876	Introduction of Milk chocolate

The Production and Processing of Chocolate

Chocolate begins with the cocoa tree, Theobroma cacao, a wide-branched evergreen that grows up to 7.5 m tall. The seedpods or fruit (which contain the much prized cocoa beans) grow up to 20 cm long and 10 cm thick, with a hard leathery shell. Each pod may contain as many as 40 seeds, or beans, each about 2 cm in size.

Several species of cocoa tree are cultivated in tropical regions around the world. Theobroma cacao, the principal species used for cocoa, is grown throughout the wet, lowland tropics, especially in Southeast Asia, South America, and West Africa. Trees usually bear their first fruit 4 years after they have been planted.

The cocoa pods are left to rot, or "ferment" for a few days to assist in the extraction of the beans. The extracted beans are covered with a silky membrane which has to be removed before the beans are fermented for a further time and then dried, to prevent them from going mouldy, and finally shipped to chocolate or cocoa manufacturers.

The manufacture of chocolate begins during the fermentation stage when a number of new chemicals are formed through enzymatic and other reactions inside the pods. Once the beans are delivered to the factory they are cleaned very carefully to remove any foreign matter. Beans from different sources are often blended to achieve delicate nuances of flavour. The cleaned, blended and dried beans are then roasted to develop the flavours. It is during this roasting stage that most of the flavour we recognise as chocolate is developed. Indeed, the roast beans themselves smell and taste like chocolate. When cooled, the beans are broken and the waste shells are separated in an air current. The shells are often sold as garden fertiliser these days, so you can have a chocolate smelling garden if you wish! After cooling, the broken cocoa beans, which are often termed "nibs", are ground between steel rollers. During this grinding the starchy part of the beans is ground into a fine powder; at the same time, the temperature rises and the fatty component of the beans melts. The molten fat then coats the ground cocoa solids to form a viscous liquid, (called chocolate liquor).

Next sugar and additional cocoa butter, and milk solids (for milk chocolate), are added to the chocolate liquor. These ingredients are thoroughly blended and ground to a fine paste using heavy machines, with rollers several metres wide and as much as a metre or so in diameter to crush the mixture repeatedly. All the solid particles (sugar, starch from the cocoa beans and any milk solids) are reduced to microscopic sizes to produce the smoothness typical of fine eating chocolate.

The chocolate is then "conched", a unique process that completely mixes the chocolate at high temperatures (54°–71°C) while exposing it to a blast of fresh air. The warm air carries away with it some unwanted volatile chemicals that formed in the earlier processing, probably the most important being acetic acid (vinegar). During the conching, yet more complex chemical changes take place to further develop the chocolate's delicate flavour.

The addition of vanilla or other natural or artificial flavours provides a final flavour note. Lecithin, an emulsifier derived from soya beans, is also added; the effect is to prevent the solid particles in the liquid fat from aggregating together and thus helps to control the overall thickness, or viscosity of the molten chocolate.

Finally the chocolate is tempered, a process of carefully heating, stirring, and then cooling the liquid chocolate. The chocolate is now ready for use in moulded bars and other products. Whatever the product, chocolate is probably the world's favourite flavour and truly deserves the designation Theobroma cacao, given to the tree by the Swedish botanist Carolus Linnaeus in 1728 when he classified the cacao plant as the "food of the gods."

The meaning of the word chocolate has changed throughout the centuries; originally chocolate was a drink. These days chocolate is the food made by combining the roasted ground kernel of the cacao bean with sugar and cocoa butter, the fat released when the bean is ground. Chocolate may also contain natural or artificial flavours, emulsifiers, and – in the case of milk chocolate – milk solids. The exact proportions of these ingredients determine the type of chocolate. Each country has its own definitions of chocolate. In the US, federal standards define several kinds of chocolate products. Bitter chocolate, or chocolate liquor, is the roasted ground kernel (nib) of the cacao bean; it is commonly known as baker's, or baking, chocolate; a maximum of 5 % cocoa butter may be added to aid setting into bars. When the amount of chocolate liquor (or cocoa solids) is greater than 35 percent, the product is bittersweet chocolate. A minimum of 15 percent liquor mixed with sugar and cocoa butter is sweet chocolate. A combination of at least 12 percent dry whole milk solids, sugar, cocoa butter, and at

Table of chocolate compositions

Description of Chocolate	Cocoa Solids	Added Cocoa Butter	Sugar Solids	Milk	Other major added ingredients	Carbo-hydrates	Protein	Fats
Bitter (USA)	>95%	<5%	0	0	0	30%	10%	55%
Bittersweet (USA)	35–50%	15%	50–35%	0	0	35–50%	5%	20–40%
Sweet (USA)	15%	15%	70%	0	0	60%	5%	30%
'Continental'	70%	5%	25%	0	0	25%	10%	40%
'Deluxe Dark'	35%	15%	50%	0	0	50%	5%	35%
Plain Chocolate	20%	15%	55%	0	8% butterfat and vegetable fats	60% 30%	4%	
Milk Chocolate	10%	20%	50%	15%	0–5% butterfat and vegetable fats	60%	5%	30%
Chocolate flavoured cake covering	0	0	55%	0–10%	35% vegetable fats, 10–20% Cocoa powder			

least 10 percent chocolate liquor produces milk chocolate. In Europe there is not yet any EU legislation defining different grades of chocolate. However, in England we define chocolate as containing a minimum of 35% cocoa solids, while most other European countries require a higher proportion of cocoa solids before a product may be called chocolate.

"Chocolate flavoured compounds"

Much of the 'chocolate' sold in the UK is made using cheaper fats than cocoa butter. Chocolate processors can get higher prices for cocoa butter from the cosmetics industry than they can from the food industry. This price differential means that it is economic to produce a range of 'Chocolate flavoured compounds' which contain virtually no cocoa butter. These compounds do not have the same sharp melting point as chocolate, and tend to melt at lower temperatures. They are however often easier to work with as they do not usually require the careful tempering that cocoa butter needs to obtain the correct crystal form.

Science of chocolate production – what is going on at each stage

Fermenting the pods and Roasting the beans

The complex flavours of chocolate are developed by the fermentation of the pods and by the roasting of the beans. Neither process on its own will give the full flavour of chocolate that we all know and enjoy. During the fermentation process many chemical reactions occur. A range of new chemicals (many of which still have not yet been identified) are formed from the natural sugars in the beans and the surrounding flesh. The beans are then extracted from the pods and roasted. During the roasting process, larger molecules are broken down into smaller ones which provide the characteristic flavour and smell of chocolate. These browning reactions involve over 300 different chemicals some of which are only formed during fermentation. Thus both the fermentation and the roasting are essential to give the full chocolate aroma.

Grinding to make 'Nibs' and Pressing

The cocoa beans contain between 50 and 55% fat (cocoa butter). In the beans as they are grown this fat is contained in very small pockets inside a rigid structure of carbohydrates and a little protein. The beans are hard and crunchy, and do not have a particularly pleasing texture when eaten. If the cocoa beans are to be converted into smooth chocolate, then the rigid carbohydrate strucctures must be broken down. The first stage in this process is the grinding of the beans to make 'nibs', in this process the beans are sufficiently broken up that the walls of the separate pockets of cocoa butter are broken. In the next stage of processing the 'nibs' are heated so that the cocoa butter melts, and then pressed so that the liquid butter flows out. The remaining solid is then ground further to make cocoa powder. The nib is also ground further to reduce the size of the remaining solid particles; the amound of the grinding varies between the manufacturers.

Adding sugar and cocoa butter

The 'nibs' can be heated and will just about melt and flow (they consist of more than 50% fat). The liquid fat completely surrounds the finely ground solid particles. However, the product is very bitter and cannot be eaten (by Europeans) without the addition of some sugar. As sugar is added to the hot, liquid, nibs so the proportion of fat drops and quickly there is insufficient fat to completely coat all the solid particles of both sugar and cocoa solids. At this stage the mixture ceases to flow easily, and cannot be processed. To increase the ease of processing and allow a smooth finished product it is necessary to add more fat. This fat ensures that the separate solid particles of cocoa solids and finely ground sugar are all kept separate and suspended in a continuous matrix of fat. Usually the added fat is exclusively cocoa butter, but manufacturers are increasingly using cheaper alternatives. In practice to make chocolate you need a minimum of 30% total fat content; if there is less fat than this the melted chocolate is so viscous that it will not flow properly into moulds, and it will have a dull and rough surface.

Conching

In the conching process the components of the chocolate ('nibs', cocoa butter, sugar, and emulsifiers) are mixed together, and some further rolling to reduce the size of the solid particles of proteins, carbohydrate and sugar is performed. At this stage the chocolate is a suspension of very fine solid particles in liquid fat. There are two different types of solid particle present: the added sugar crystals, and the ground up particles of carbohydrate and protein from the cocoa beans. The sizes of these particles have a major influence on the texture of the final chocolate. In 'Continental' chocolate the particles are ground down to extremely small sizes (ca. 0.002 mm) while in England we like our chocolate with coarser particles (about 0.01 mm). Although in all cases the particles are rather smaller than we could tell from biting them, the texture of the molten chocolate is strongly influenced by the size of the solid suspended in it.

Many liquids contain a suspension of small solids, we call such liquids 'emulsions'. The solid particles are prevented from sticking to each other by the use of 'emulsifiers'; these are molecules one end of which sticks to the surface of a solid particle (so that the particles become coated with a layer of these molecules). The end of the molecule not attached to the solid particles dangles out in the liquid, and is designed so that it wants to be surrounded by the liquid. This process of coating solid particles to stabilise them in suspension is how soaps and detergents work (see Chapter 2). In chocolate it is necessary to add the emulsifying molecules as they do not occur in the cocoa bean. The molecule lecithin, which is extracted from soya beans, is usually used for this purpose in chocolate manufacture.

The viscosity (or runniness, or thickness) of an emulsion is determined not only by the viscosity of the liquid, but also, by the amount, and importantly the size, of the solid particles in the emulsion. The smaller the particles, the thicker the resulting emulsion will be. This thickening of the emulsion when the solid particles are very small is the reason for the different texture of continental and English chocolate. When the chocolate melts in the mouth, English chocolate flows rather more quickly around the mouth, so we get the 'feel' and taste of it more quickly; however the continental chocolate is more viscous or thicker, and it stays around in the mouth much longer giving a more lingering taste.

During the conching the initial unstable mixture of solid sugar crystals, cocoa powder, molten cocoa butter and emulsifiers is gradually converted into a stable emulsion. It is very important never to overheat chocolate since the emulsifiers can be stripped from the sugar crystals at temperatures above 55 °C and they will usually degrade (break up) at temperatures above 75 °C. Thus heating chocolate above 55 °C will tend to allow the sugar crystals to aggregate, and can give the chocolate a coarse texture, unless it is carefully stirred to repeat the conching. Heating above 75 – 80 °C will irreversibly destroy the emulsifying molecules and will mean that only a coarse chocolate can be made.

Crystal formation and Tempering

The final stage in the production of chocolate is to cool the molten chocolate down and crystallize the cocoa butter. This is perhaps the trickiest stage. Cocoa butter can crystallize in any of six different forms! Only one is right for chocolate. Many materials can exist in several crystal forms which have very different properties. A good example is carbon. Two of the forms of crystalline carbon are graphite (used as the lead in pencils) which is a soft, black, material; and diamonds which are extremely hard and transparent. Similarly, the different crystal forms of cocoa butter vary in their hardness, and in their melting temperatures. For chocolate we need the crystals to be hard enough to hold the chocolate together in a strong, but breakable bar. If the crystals were too strong we could not bite into a chocolate without breaking our teeth! We also need the melting point of the crystals to lie above room temperature but below the temperature inside our mouths, so that the chocolate really does melt in the mouth. The final requirement is for small crystals; if the crystals were larger than the sizes of the other solid particles then the chocolate would taste gritty, and it would not be possible to get a smooth, glossy finsih the solid bars.

The particular form that cocoa butter crystallizes in is determined mainly by the temperature at which it starts to crystallize. The desirable form will dominate if the crystallization temperature is between about 18 and 25 °C. Once a crystal starts to grow it will continue to grow in the same form, so it is important that crystals are prevented from starting to grow (nucleate) at high temperatures. Also if the crystals are to be kept small then very many such 'nuclei' need to be formed in the narrow temperature interval where the desirable crystal will be dominant. Growth outside the narrow temperature range will lead to increasing numbers of less desirable crystals, and to rather tough and 'chewy' chocolate.

The tempering process involves first heating the chocolate to a high enough temperature (above 44 °C) and for a long enough time to melt all existing cocoa butter crystals, but without overheating to destroy the carefully made emulsion. Next, the chocolate is carefully cooled (to below 28 °C); at this temperature cocoa butter crystals of at least two of the different forms can grow. As these crystals are being formed the molten chocolate is stirred. The stirring action breaks these small, growing, crystals into small pieces, thus increasing their number. As more crystals are formed and broken up so the viscosity of the molten chocolate increases. Just before the chocolate can set the temperature is raised to 31 °C. At this higher temperature the undesirable, lower melting point, cocoa butter crystals will melt, leaving only the desirable form that has a melting point of about 33 °C stable. With the reduction in the number of crystals the chocolate once again becomes fluid and is ready for use. Now the chocolate is ready, it may be poured into moulds or used to coat fillings and then cooled in a fridge where the crystal nuclei quickly grow, still in the desirable form, to solidify the chocolate.

The cocoa butter crystals are not always completely stable, especially if they are heated above about 20 °C. The cocoa butter may slowly migrate from the crystals to the surface of the chocolate where it recrystallizes as a white bloom.

Making Chocolates

There are many ways to make chocolates, and chocolate confectionery. Most large scale processes for making chocolates involve passing the cold solid centres through a curtain of molten chocolate which coats them. Liquid centres are often made by taking advantage of a enzyme reaction which can convert initially solid centres into liquid ones. In this process the centres are made using sugar, and a little water, and an enzyme is added which converts the sucrose in the sugar to the more soluble glucose and fructose. The result is a change from a solid centre to a creamy one over a period of up to two months at 18 °C.

The small chocolatier usually coats his centres by hand, sometimes by dipping into the molten chocolate, and sometimes using moulds that are first lined with a thin coating of chocolate. A top coat is added after the filling. Some liqueur chocolates are produced by freezing the liqueur to make a solid which can be coated with chocolate. However it is usually neccessary first to coat them with a layer of sugar, since the alcohol in the liqueur can dissolve the cocoa butter crystals and the chocolates would then dissolve.

Dr. Pringle's Chocolate Oranges

Dr Sue Pringle is one the best exponents of the communication of science to the public. A few years ago she organised a series of public lectures in Bristol and convinced me that I should contribute with a talk on "The Physics of a Black Forest Gateau". That lecture started me thinking more deeply about the science of food and indirectly led to the publication of this book!

Over the years Sue became a very good friend and we started giving dayschools together – one of the most successful being on Chocolate. However, as Sue will readily admit, she is not one of the World's good cooks. She learnt the hard way that "breadcrumbed chicken breasts" sold in a supermarket were not cooked!

A few years ago by a series of odd coincidences I was a guest, together with Sue, at a dinner party held by a Michelin Star chef, who also happened to be Sue's next door neighbour. The idea had been that Ramon, the chef, would provide the starter and main course, but Sue would provide a dessert. After several false starts, Sue determined to make chocolate coated orange segments. The problem was that Sue had never before tried to work with chocolate and had no idea how to coat anything with chocolate.

So when the time came for dessert we were invited next door to Sue's kitchen where she opened her fridge to reveal a set of strange, knobbly, sausage shaped, brown objects hanging by red cotton threads from the shelves. The bottom of the fridge was covered with chocolate. It turned out that Sue had melted the chocolate in a saucepan, paying no attention to the temperature and then stirred the orange segments into the molten chocolate. She then took a needle and thread and sewed a length of cotton into each segment before hanging them in her fridge to set.

We greatly enjoyed the spectacle, but not, I am sad to say, the chocolate oranges – so Ramon quickly whipped up some soufflés instead!

Savoury Mexican Chocolate Sauce or Mole

The rich sauces made by combining chillies, tomatoes and nuts with chocolate and thickened using ground roasted seeds, have one of the most remarkable flavours in the world. These sauces were developed centuries ago for the rulers of South America long before the Spanish arrived. They survive today in Mexican cuisine, but are rarely seen outside the Americas. It is not always easy to obtain the ideal ingredients in Europe, but some of the better supermarkets are starting to stock a selection of dried chillies and they are often available by mail order. However, these chillies are always expensive, so if you are in North America it is best to stock up and bring a large supply home – the dried chillies will keep very well in a refrigerator for many months.

Since it takes a long time to make this sauce (about 4 hours or more) I find it very useful to make this sauce in large quantities and freeze in meal sized portions ready to use as and when required.

Recipe for "Mole"

Ingredients
4 dried ancho chillies
2 dried mulato chillies
1 dried pasilla chilli
Or 6 fresh chillies (if possible use several different varieties) about 3 oz total weight
$1^1/_2$ tablespoons sesame seeds
$^1/_4$ pint vegetable oil
2 tablespoons shelled peanuts – freshly roasted, if possible
2 tablespoons raisins
$^1/_4$ medium onion
1 clove garlic
6 tortilla chips
1 thick slice of stale white bread
1 small tomato (roasted or boiled)
1 small tin tomatoes
$^3/_4$ ounce of unsweetened chocolate (if you can't get unsweetened use choco late with *at least* 75 % cocoa solids)
$^1/_2$ teaspoon oregano
$^1/_4$ teaspoon thyme
1 bay leaf
8 peppercorns
3 cloves
1 inch cinnamon stick
$2^1/_2$ pints chicken stock

Begin by preparing the chillies. If you are using dried chillies tear off the ends and empty them of seeds, then tear the chillies into small pieces. Now fry the chillies, a few pieces at a time in hot oil for a few minutes. As they are fried the chilli pieces will puff up to regain a smooth texture and may also regain some of their original red colour. At the same time *very* punget odours will be generated – so make sure your extractor fan is turned full on and the windows are wide open! Once the chillies have been fried put them in a bowl and cover with boiling water. Cover and leave the chillies to soak for an hour. Drain the chillies, keeping the now dark liquid, and puree them in a food blender using just enough of the liquid to keep the blades moving. If you are using fresh chillies, begin by deseeding the chillies and cut them into pieces. Then fry them for a few minutes until they just start to brown. Then soak and puree as for the dried chillies. ****Remember that these chillies can be very hot, as you have been handling them be very careful where you put your hands. For example, if you should wipe your eyes you may end up crying for an hour or more!****

Toast the sesame seeds until they are golden brown – put aside. Fry, in the same pan adding more oil if needed, the peanuts, raisins, onion, tortilla chips and the bread. Mix all the ingredients (except the chicken broth and pureed chillies) together and stir well. Blend this mixture with about half the chicken broth in a food processor (you may need to do this in two batches depending on the size of your blender) and then puree this further in a blender, adding more stock if needed. Heat the mixture in a saucepan stirring all the time until it thickens, simmer for at least 20 minutes. Add the pureed chillies to the mixture in small amounts, tasting after each addition to ensure you do not make the sauce too hot for your taste. Leave the sauce to simmer for at least an hour, add more stock if necessary (the sauce should be very thick as it is meant to coat the meat you are eating with it).

Use the sauce with chicken breasts that have been lightly sautéed, or with turkey, or with stuffed peppers, etc. The sauce has one of the most glorious flavours I know and can be eaten with just about anything you like. If you ever get the chance visit a really good, authentic, Mexican Restaurant to taste just how good this sauce can be. I have only been to two places that really make it well, one in San Diego and the other in Dallas.

Truffles

Truffles are probably the easiest sort of chocolates to make at home, but can taste and look so good that they impress far more than is really deserved considering the ease with which they can be made. There are many possible variations, so I will begin with a basic recipe and then just list some of the variations for you to try for yourself. Remember that you can invent your own recipes so that you can make truffles to suit any occasion, or to fit in with any particular taste, etc.

Basic recipe

Ingredients
200 g Dark Chocolate (use a good chocolate with at least 50 % Cocoa solids – preferable 70 % cocoa solids)
100 ml double cream
Cocoa powder to coat the truffles

Methods

Break the chocolate into small pieces and melt in a bowl over hot water. Take care not to overheat the chocolate (i.e. keep the temperature below 45 °C). There is a real risk that the "emulsifying" agents in the chocolate may be destroyed at higher temperatures, this will make the chocolate "curdle". Stir the melted chocolate, which should have a good sheen and be completely free from any lumps. Set the chocolate aside and let it cool a little (try to keep the temperature at around 30 °C – i.e. just warm to the touch). Lightly whip the cream and gently stir it into the chocolate to form a uniform texture. Leave the mixture to cool in a cold room or in the refrigerator for about half an hour when it should be very stiff, but not fully solid. If it does set solid then you can simply melt it over hot water and let it cool again. Now you have to form the mixture into small balls – if you try to do this with your hands you will melt the mixture and have very messy fingers, so try to use a pair of cold, metal spoons. Scoop a spoonful of mixture up with one spoon and carefully shape and remove it with the other dropping it onto a plate of cocoa powder; roll the truffle in the cocoa powder and put in a paper case. Once all the truffles have been made put them in the refrigerator for an hour or so to set fully then allow them to warm up to room temperature just before serving.

Recipe variations for truffles

There are very many possible variations.

First you can adjust the texture of the truffles by whisking the mixture just before you shape them – use a power whisk on a slow speed to incorporate air in the mixture, which will become significantly lighter in colour. Beating in air in this way produces a lighter truffle.

Secondly, you can add flavour to the truffles. You can use any flavour you want; for example, adding a few drops of orange oil will produce orange truffles. Other possibilities include coffee, and nut flavoured truffles. The recipe given above is very robust and you can vary the proportions and include up to 100 g of any solids such as nuts with no problem.

Thirdly, you can make alcoholic truffles. All you do is add the brandy, or any other flavour, to the melted chocolate at the beginning. You can easily incorporate up to 50 ml of alcohol in this way.

Finally, you can change the outside finish of the truffles. For example you can coat them with a little melted chocolate to give a shiny chocolate coated truffle, or you can roll them in grated nuts such as almonds, etc.

Easter Eggs

It is very simple to make your own Easter eggs. All you need is some good quality chocolate and the Easter egg moulds (these can be purchased in any good cooks shop). You simple melt and temper the chocolate and then paint it in layers over the inside of the mould. Once the chocolate has set it will come away from the mould easily and you can repeat to make a second half egg.

Ingredients

200 g Dark Chocolate (use a good chocolate with at least 50 % Cocoa solids – preferably 70 % cocoa solids)

Methods

Prepare the moulds according to the manufacturers instructions – make sure they are very clean and dry.

Melt the chocolate heating it (over a basin of hot water) to just above 44 °C – it will form a thick glossy liquid. Next cool the chocolate to 28 °C over a bowl of cold water beating it all the time. If you do not have a thermometer you need to stop as soon as the chocolate starts to thicken. Now reheat the chocolate over the hot water to 31 °C when it should have a glossy sheen. Now the chocolate is ready to use. With a fine paintbrush coat the inside of the mould with a thin layer of chocolate, allow it to begin to set (which should only take a few minutes) and then paint on a thicker layer (up to a millimetre) and allow that to start to set before painting on another layer. Repeat until the desired thickness is achieved. Leave the mould in a cool place (but not in the refrigerator as this is too cold and will encourage the formation of lower melting point crystals) until the chocolate is fully solidified and then gently prise the egg from the mould. Repeat to make a second half egg.

When you have made both halves of the egg you can glue them together simply by painting a little melted chocolate on the edges and pressing the halves together. If you want to impress the person you are giving the egg then you can place a small gift or some other chocolates inside the egg before sealing it up.

An Experiment to try for yourself

Plastic Chocolate

The process of making chocolates can be quite cumbersome, especially the tempering process that is needed to ensure the cocoa butter ends up in the right crystal form with the required melting temperature. One of my old colleagues, Malcolm Mackley, now at Cambridge University, applied some of his understanding of the flow properties of polymeric fluids, to the processing of chocolate. As he admits himself, if he had known about the complexities of the processing of chocolate he would never have tried this simple experiment. However, he has invented a whole new way of processing chocolate and making it into shapes that otherwise would be very difficult if not impossible.

What Malcolm discovered is that if solid chocolate is pushed through a hole (extruded) it becomes a malleable plastic material that can be moulded to any desired shape and then, after some short time it re-solidifies in the same crystal form it started out.

You can try this out for yourselves at home if you have a mincing machine (preferably a power machine, although a hand operated one will work). Alternatively, a pasta machine set up to make spaghetti can be used. Simply break up a bar of chocolate into the hopper of the mincing machine and then start it up using the coarsest set of grinders available. The chocolate will be pushed out through the holes at the end of the mincing machine in continuous strands. These strands will be quite flexible and you can tie them in knots or weave them together as you wish. However, if you leave them to stand for an hour or so they set back as hard as the chocolate they started from.

With a little experimentation to adjust the speed of the mincing machine and a little practice you can perform this trick with more or less any chocolate and make up all sort of interesting shapes to impress your friends and relatives.

Weights and Measurements

Throughout the text I have exclusively used the System International (SI) or metric units of measurements. These are the units used in all science laboratories throughout the world and leave no room for ambiguity.

However, as I realise that in some kitchens people may prefer to use the older imperial (or US) measurements when cooking I have included below some tables for quick reference to allow conversion between the different units in common usage.

These conversions are approximate and are rounded to the nearest convenient number. However, any errors should be unimportant in cooking as all the recipes given in this book are fairly robust and do not rely on exact quantities.

Measurements of Weight

Metric	Imperial or US
10 grams	$^1/_2$ oz
20 grams	$^3/_4$ oz
25 grams	1 oz
50 grams	2 oz
100 grams	3 oz
150 grams	5 oz
200 grams	6 oz
250 grams	9 oz
300 grams	10 oz
350 grams	12 oz
400 grams	14 oz
450 grams	1 lb
500 grams	1 lb 2 oz
1 kg	2 lb 4 oz

Measurement of Volume

Metric	Imperial or US		
5 mL			1 teaspoon
10 mL			2 teaspoons
25 mL			5 teaspoons
50 mL		1/5 cup	10 teaspoons
100 mL	1/5 pint	2/5 cup	
150 mL	1/3 pint	2/3 cup	
250 mL	1/2 pint	1 cup	
500 mL	1 pint	2 cup	

Measurement of Temperature

Celsius	Farenheit
30 °C	85 °F
40 °C	105 °F
50 °C	120 °F
60 °C	140 °F
70 °C	160 °F
80 °C	175 °F
90 °C	195 °F
100 °C	212 °F
120 °C	250 °F
140 °C	285 °F
160 °C	320 °F
180 °C	360 °F
200 °C	390 °F
225 °C	440 °F
250 °C	480 °F

Measurement of distance

Metric	Imperial or US
5 mm	1/5 in
10 mm	2/5 in
25 mm	1 in
50 mm	2 in
100 mm	4 in
200 mm	7 3/4 in
500 mm	1 foot 7 1/2 in
1 metre	3 feet 3 1/2 in

Measurement of dry goods by volume, rather than weight

In many American households it is common to use volumes, rather than weights to measure out dry goods. Since the volume of a given weight of different substances (sugar, flour, etc.) depends on the density of the substance, the conversion between weight and volume depends on what is being measured. The tables below provide a range of examples for most common foodstuffs.

Butter, Shortening, Cheese, and Other Solid Fats

1 tablespoon	$^1/_8$ stick	$^1/_2$ ounce	15 grams
2 tablespoons	$^1/_4$ stick	1 ounce	30 grams
4 tablespoons ($^1/_4$ cup)	$^1/_2$ stick	2 ounces	60 grams
8 tablespoons ($^1/_2$ cup)	1 stick	4 ounces ($^1/_4$ pound)	115 grams
16 tablespoons (1 cup)	2 sticks	8 ounces ($^1/_2$ pound)	225 grams
32 tablespoons (2 cups)	4 sticks	16 ounces (1 pound)	450 grams
			(500 grams =
			$^1/_2$ kilogram)
50 grams	3 $^1/_3$ tablespoon		
100 grams	$^1/_2$ cup minus 1 tablesoon		

Flours (unsifted)

1 tablespoon	$^1/_4$ ounce	8.75 grams
$^1/_4$ cup (4 tablespoons)	$1^1/_4$ ounces	35 grams
$^1/_3$ cup (5 tablespoons)	$1^1/_2$ ounces	45 grams
$^1/_2$ cup	$2^1/_2$ ounces	70 grams
$^2/_3$ cup	$3^1/_4$ ounces	90 grams
$^3/_4$ cup	$3^1/_2$ ounces	105 grams
1 cup	5 ounces	140 grams
$1^1/_2$ cups	$7^1/_2$ ounces	210 grams
2 cups	10 ounces	280 grams
$3^1/_2$ cups	16 ounces (1 pound)	490 grams
100 grams	$^3/_4$ cups minus $^1/_2$ tablespoon	
250 grams	2 cups minus 3 tablespoons	
400 grams	3 cups minus 2 tablespoons	
500 grams	$3^1/_2$ cups plus 1 tablespoon	

Granulated Sugar

1 teaspoon	$^1/_6$ ounce	5 grams
1 tablespoon	$^1/_2$ ounce	15 grams
$^1/_4$ cup (4 tablespoons)	$1^3/_4$ ounces	60 grams
$^1/_3$ cup (5 tablespoons)	$2^1/_4$ ounces	75 grams
$^1/_2$ cup	$3^1/_2$ ounces	100 grams
$^2/_3$ cup	$4^1/_2$ ounces	130 grams
$^3/_4$ cup	5 ounces	150 grams
1 cup	7 ounces	200 grams
$1^1/_2$ cups	$9^1/_2$ ounces	300 grams
2 cups	$13^1/_2$ ounces	400 grams
100 grams	$^1/_2$ cup	
250 grams	1 cup plus 3 tablespoons plus 1 tea spoon	
400 grams	2 cups	
500 grams	$2^1/_2$ cups	

Other Equivalents

Bread crumbs (4 sandwiches)			
Dry	³/₄ cup	4 ounces	115 grams
fresh	2 cups	4 ounces	115 grams
Brown sugar	1¹/₂ cups	1 pound	450 grams
Confectioners' sugar	4 cups	1 pound	450 grams
Egg whites			
1	2 tablespoons		
8	1 cup		
Egg yolks			
1	1 tablespoon		
16	1 cup		
Fruit, dried and pitted			
Plumped	2²/₃ cups	1 pound	450 grams
Cooked and puréed	2¹/₃ cups	1 pound	450 grams
Fruits, fresh, such as apples			
Raw and sliced	3 cups	1 pound	450 grams
Cooked and chopped	1¹/₃ cups	1 pound	450 grams
puréed	1¹/₄ cups	1 pound	450 grams
Nuts, chopped	³/₄ cup	4 ounces	115 grams
ground	1 cup loosely packed	4 ounces	115 grams

Vegetables

Carrots and other roots			
Sliced	3 cups	1 pound	450 grams
puréed	1¹/₃ cups	1 pound	450 grams
Onions sliced or chopped	3 cups	1 pound	450 grams
Potatoes, raw, sliced or chopped	3 cups	1 pound	450 grams
Spinach and other leafy greens	1¹/₂ cup	1 pound	450 grams

Liquid and Dry Measure Equivalents (US)

Liquid and Dry Measure Equivalents (US)		Liquid	Dry
2 tablespoons	1 ounce	25 mL	30 grams
1 cup	¹/₄ quart	250 mL	225 grams
2 cups	1 pint	500 mL	450 grams
4 cups	32 ounces	1000 ml (1 L)	
4 quarts	1 gallon	3.75 L	

Glossary of Terms

Acids Chemicals that react, often strongly, with a range of materials and which neutralise alkalis. Edible acids, such as citric and acetic acids are generally sour tasting. Acids turn vegetable blue colours to red colours.

Aldehydes A class of chemicals related to alcohols (Aldehydes are formed by removing two hydrogen atoms from alcohols). Many aldehydes are strong flavour molecules.

Alkalis Chemicals such as caustic soda that neutralise acids. Alkalis turn vegetables red colours blue, yellows brown and purples green.

Amylopectin The branched molecules in starch which consists of many glucose rings joined together to form long molecules with a number of short side branches.

Amylose The linear molecules in starch which consists of many glucose rings joined together to form long molecules with no side branches.

Atoms The small particles which are the basic building blocks of all matter.

Bound water Water that is "linked to" or "tied up within" other matter, usually proteins, so that it is no longer in a liquid form and is unable to flow freely.

Carbon dioxide A simple gas the molecules of which are made from two atoms of oxygen joined to one atom of carbon. Carbon dioxide is generated by yeast and baking powder to provide the rising action in cakes and breads.

Casein The collective name for a group of proteins found in milk which are precipitated in an alkaline environment.

Cellulose A long carbohydrate molecule made from many glucose rings joined together. Cellulose differs from Amylose in the ways in which the glucose rings are fixed to one another.

Colloid A dispersion of one (normally liquid) substance (in the form of small drops) within another liquid substance. The drops of the dispersed liquid phase are usually less than 1/1000 of a millimetre in size. In cookery, mayonnaises are examples of colloids where very small oil droplets are dispersed in a watery phase.

Crystal A highly ordered solid in which the molecules are arranged on a regular repeating lattice.

Crystallization The act of growing a crystal of a substance from a disordered phase. The disordered phase may be either the molten substance itself (for example when growing ice crystals from water) or a solution of the substance (for example when salt crystals grow as water evaporates from a saline solution).

Deglaze The process of removing all the "tasty" browned bits from a pan in which some meat has been cooked. To deglaze a pan add some boiling water or stock, and scrape around the pan with a spoon or spatula until all the dark brown material is dissolved in the water and then pour off the liquid and keep it for further use.

Degrade, degradation When a long molecule is heated or stretched too much it is sometimes unable to take the strain and breaks into two smaller molecules. This process of breaking large molecules into smaller pieces by mechanical force, or heat, is called degradation.

Denaturation, denature, denatured In their natural state, proteins adopt a specific conformation or shape. All molecules of the same protein have identical shapes. It is this shape that gives them their biological function. If a protein loses its natural shape, through heating, or other means it is said to be denatured.

Density The ratio of an object's weight to its volume. Every material has its own density; metals, such as iron and lead which may seem heavy have a high density, while other "lighter" materials such as wood and plastics have lower density.

Di-sulphide bridge One of the various types of chemical bonds that occur within protein molecules and help to hold them in their "natural" shape.

Emulsion A dispersion of small drops of one substance inside another. Butter is an example of an emulsion of drops of water suspended in a solid fat. Mayonnaise is an example of an emulsion of oil drops suspended in water.

Enzyme Enzymes are naturally occurring chemicals that help to promote biochemical reactions. Nearly all biological processes are controlled by the production, or suppression of the appropriate enzyme.

Exponentially, exponential increase These are mathematical terms describing series that increase very rapidly. If you place one penny on one square of a chessboard, two on the second, four on the third, eight on the fourth, sixteen on the fifth, and so on; doubling the number of pennies on each successive square, then you could say the number of pennies on each square is increasing exponentially.

Fibril Literally, a small fibre. When describing muscle structure, fibrils are the smallest units that the individual muscle fibres can be seen to be composed of.

Gliaden One of the proteins in wheat flour which, together with glutenin, can form gluten. Gliaden is distinguished from glutenin by being soluble in alcohol.

Gluten A protein complex (a combination of two or more proteins) formed by mechanical deformation of mixtures of the wheat flour proteins gliaden and glutenin with a little water. Gluten is generally formed when flour doughs are kneaded.

Glutenin One of the proteins in wheat flour which, together with gliaden, can form gluten. Glutenin is distinguished from gliaden by being insoluble in alcohol.

Hydration The act of adding water to a substance – usually referring to situations where the water either reacts chemically with the substance, or is absorbed by it.

Hydrogen bond One of the various types of chemical bonds that occur within protein molecules and help to hold them in their "natural" shape.

Hydrolysis A chemical reaction where a molecule is split into two separate, smaller, molecules by the addition of a water molecule.

Hydrophilic Has a liking for, or is attracted to, water. Alcohol mixes easily with water and is hydrophilic. The opposite of hydrophobic.

Hydrophobic Dislikes, or is repelled from, water. Oils do not mix with water and are hydrophobic. The opposite of hydrophilic.

Inter-protein bonds Chemical or physical bonds between different protein molecules. When many protein molecules are bonded together they form gels.

Intra-protein bonds Chemical of physical bonds that occur within a single protein molecule. These are the bonds that help to hold it in its natural shape.

Ion, ionic bond Ions are atoms that have either gained, or lost, one or more electrons so that they are electrically charged. Ions can form bridges, ionic bonds, between (or within) molecules joining together charged regions.

Maillard reactions A complex set of chemical reactions that occur between sugars (including polysaccharides and small sugar molecules) and proteins. The reactions only occur at an appreciable rate at high temperatures (above about 140 °C). Many of the reaction products are small, flavour molecules – providing, for example, the "meaty" flavour of cooked meats. The Maillard reactions are also sometimes called the browning reactions as they occur when food is browned in the oven, or under the grill.

Metabolise The action of enzymes to promote the reduction of complex food molecules into smaller more useful molecules.

Micelle A small aggregate of molecules such as soaps, or casein proteins. Micelles are relatively stable in colloids.

Molecule A group of atoms chemically bonded together.

Nucleate, Nucleation When a crystal, or a bubble, starts to grow, the very first stage is the formation of a nucleus (nucleation). If the nucleus is too small, then the bubble will collapse, or the crystal melt. The nucleus has to have a sufficient size before it becomes stable and the crystal, or bubble, can grow.

Ordered When describing molecules in the solid state, we may say they are "ordered" or "disordered". In an ordered solid, the location and orientation of one molecule influences that of its neighbours, while in a disordered solid it does not. A very well ordered solid would be a crystal where each molecule occupies an identical site on a repeating lattice.

Oxidation Chemical reaction where oxygen is added to a substance. In fats, oxidation makes them become rancid.

Polysaccharides Long polymer molecules made by joining together many sugar molecules. There are several different sugars and each can be joined through different carbon atoms in their rings so that there is an infinite set of possible polysaccharides.

Protein Long polymer molecules made by joining together many amino acids. There are several amino acids and these can be joined in any order so that there is an infinite set of possible proteins.

Rennin An enzyme from the lining of calves' stomachs that prevents the casein in milk from forming stable micelles and results in the formation of a curd.

Shear-thinning Many complex fluids have odd properties when they flow. If the viscosity (the resistance to flow) reduces as the rate at which the liquid flows increases it is said to be shear-thinning.

Starches A complex of carbohydrates, mostly amylose and amylopectin.

Starch granule A granule made up from concentric layers of amylose and amylopectin molecules.

Surfactant A molecule, usually with one hydrophobic end and one hydrophilic end, that modifies, or stabilises a surface, in particular detergents and soaps which are able to stabilise oil-water emulsions.

Thixotropic Many complex fluids have odd properties when they flow. If the viscosity (the resistance to flow) decreases as it is sheared, or spread out it is said to be thixotropic.

Trigeminal sense The sense that allows us to "taste" chillis, etc.

Umami The flavour sensation provided by, for example, mono-sodium glutamate (MSG). The Umami sensation is best described as a rounding off, or enhancing effect. It is present in many foods, especially tomato purees and parmesan cheese.

Unleavened Breads made with no yeast, so that they do not "rise" and contain lots of bubbles.

Viscosity The ratio of the pressure applied to a liquid flowing through some apparatus and the rate at which it flows. The resistance to flow.

Bibliography

I have drawn on a variety of texts to assist in the preparation of this book; all are listed below, together with several other useful and relevant books. I have divided this bibliography into a few sections each covering different areas. I have tried to give a short statement of what can be found in each of these references.

Food Science Texts

Food Science; H. Charley, John Wiley, New York, 1982.
This is a comprehensive textbook for students of food science. It provides a wealth of reference material. The language may not always be easy to follow for those without some background in the sciences.

Foods: Experimental Perspectives; Margaret McWilliams, Macmillan, New York, 2nd Edition, 1993.
Margaret McWilliams has written several books on food science, this book is intended for serious students, but achieves a highly readable balance between theoretical aspects and experiments (and questions) to illustrate the basic principles.

Food Chemistry; H.-D. Belitz and W. Grosch, 2nd Edition, Springer, Berlin, 1999.
This book has only recently been published in English. I have found it to be a very thorough account of the detailed chemical reactions that occur during cooking, etc. It does assume a good deal of background chemical knowledge, but remains quite readable for a Physicist such as myself.

"Popular" food books with a science element

On Food and Cooking; Harold McGee, Unwin Hyman, London, 1986.
Harold McGee's first book is widely recognised as the best and most readable book that explains, in simple terms, the science behind our food and cooking.

The Cook Book Decoder; Arthur E. Grosser, Beaufort Books, New York, 1981.
Art Grosser is a Professor of Chemistry at Magill University. His little book is full of witty and clear explanations of the chemistry of cooking.

The Curious Cook; Harold McGee, North Point Press, San Francisco, 1990.
In his second food book, Harold McGee explores a few phenomena in greater detail. The chapter on mayonnaise is particularly enlightening.

The Epicurean Laboratory; Tina Seelig, W. H. Freeman, New York, 1991.
This short, well illustrated book explores a variety of areas where scientific explanations can be illustrated by simple recipes.

Cookwise; Shirley O. Corriher, William Morrow, New York, 1997.
Shirley is a great cookery teacher and loves to understand everything she does in simple scientific terms. Her book brings out her own enthusiasm. Throughout the text, Shirley tries to ensure that the reader understands why the recipes have their particular ingredients and instructions; the understanding then helps to ensure perfect results every time.

Blanc Mange; Raymond Blanc, BBC Books, London, 1994.
This book was written to accompany a television series where Raymond Blanc explored some aspects of chemistry behind his own cooking style. The "science" in the television series was largely explained by Prof Nicholas Kurti, and his influence can also be seen in the book.

The French Cookie Book; Bruce Healy, William Morrow, New York, 1994.
Bruce Healy has written a mammoth book on the art of the patissier, in many places, such as the rising of sponge cakes, the formation of foams, and the tempering of chocolate, he provides detailed scientific explanations of the traditional processes.

Lutefisk, Rakefisk and Herring in Norwegian Tradition; Astri Riddervold, Novus Press, Oslo, 1990.
This short book covers more than anybody could ever want to know about the Norwegian tradition of dried fish and the methods of re-constituting it. I found it in valuable when trying to understand the ins and outs of Lutefisk.

General cookery books

The Joy of Cooking; Irma S. Rombauer and Marion Rombauer Becker, Signet, New York, 1931.
This American book provides clear and easy to follow recipes for just about everything. The only problem I have with it is the use of US measurements!

Authentic Mexican; Rick Bayless, William Morrow, New York, 1987.

As a lover of good Mexican food I have found this book to be indispensable as it provides authentic tasting recipes in a clear fashion.

Good Housekeeping Cookery Book; Ebury Press, London, revised edition, 1983.

This is the traditional British cookery book – the one my mother used. However, the newer, updated editions provide good basic recipes and I keep my copy in the kitchen as a useful reference source if I forget a simple recipe.

Cordon Bleu Cookery; Rosemary Hume and Muriel Downes, Octopus, London, 1975.

This is the book that sparked my interest in cooking. Shortly after completing my doctorate, my partner bought it to encourage me to take up cookery as a hobby. I have tried and enjoyed just about every recipe in the book. However, I have to admit that some of the recipes could do with more explanation and in some cases modification. Indeed, it was by working out just how these recipes were supposed to work that I started to look into the science behind cookery.

Fruits of the Sea; Rick Stein, BBC Books, 1998.

Science texts

Since I work in a University Physics Department, I have access to large libraries full of advanced textbooks on all aspects of bio-chemistry, chemistry, biology and physics. I have of course consulted many such texts when thinking about the science of the kitchen. However, very few, if any would be available, or accessible, to the general reader, so I do not think it appropriate to list them all here. Instead, I recommend a few well written books intended for a general audience as well as a couple of approachable texts that while written for undergraduates can largely be followed by any reasonable educated person.

The New World of Mr Tompkins; George Gamow, Russell Stannard, Cambridge University Press, Cambridge, 1999.

Molecules at an Exhibition; John Emsley, Oxford University Press, Oxford, 1998.

Chemical Magic; Leonard A. Ford, E. Winston Grundmeier, Dover Publications, 1993.

Chemistry: Molecules, Matter, and Change; Peter Atkins, Loretta Jones, W. H. Freeman, 1997.

Schrodinger's Kittens; John Gribbin, Phoenix Paperbacks, 1996.

The Flying Circus of Physics With Answers; Jearl Walker John Wiley 1987 .

Principles of Modern Chemistry; David Oxtoby, H. P. Gillis, Norman H. Nachtrieb, Saunders College Publishing, 1998

Conduction of Heat in Solids; H. S. Carslaw and J. C. Jaeger, Clarendon Press, Oxford, 1947.

Index

Printing: Mercedes-Druck, Berlin
Binding: Stürtz AG, Würzburg